PROFESSIONAL
PLUMBING
TECHNIQUES
ILLUSTRATED & SIMPLIFIED

PROFESSIONAL
PLUMBING
TECHNIQUES
ILLUSTRATED & SIMPLIFIED
BY ARTHUR J. SMITH

 TAB BOOKS Inc.

BLUE RIDGE SUMMIT, PA 17214

FIRST EDITION

SECOND PRINTING

Printed in the United States of America

Reproduction or publication of the content in any manner, without express per-
mission of the publisher, is prohibited. No liability is assumed with respect to
the use of the information herein.

Copyright © 1984 by TAB BOOKS Inc.

Library of Congress Cataloging in Publication Data

Smith, Arthur J.
    Professional plumbing techniques—illustrated
and simplified.

    Includes index.
    1. Plumbing.   I. Title.
TH6122.S55   1984   696'.1   84-8883
ISBN 0-8306-0763-3

Cover illustration by Sandy Rook.

Contents

PROFESSIONAL
PLUMBING
TECHNIQUES
ILLUSTRATED & SIMPLIFIED

ACID WASTE DISPOSAL

Acid wastes harmful to house plumbing must be rendered innocuous by dilution or neutralization. Treatment tanks must be sized according to the amount to be diluted or treated. All piping carrying acids must be approved materials for this use.

All vents from acid wastes must be separated from the house plumbing and be approved for acid venting.

Clear water wastes from plumbing fixtures may enter the dilution tank provided the material for waste and vents are acid resisting.

Outside sewers may be vitrified clay with approved joining materials, cast iron, or other approved materials allowed by code.

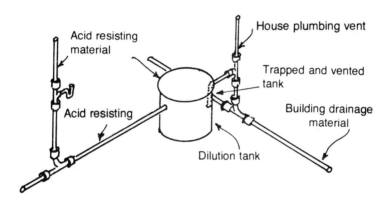

Do not connect acid wastes and vents to any house plumbing until treated. Do not back vent acid waste vents into the house venting system at any time. Check for research approved glass or plastic materials.

Acid wastes may be of high silicon iron, lead, or approved ceramic glazed vitrified clay. Vents may be any of above and extra heavy cast iron. (No clay inside buildings.)

THE ANGLE MEASUREMENT CHART

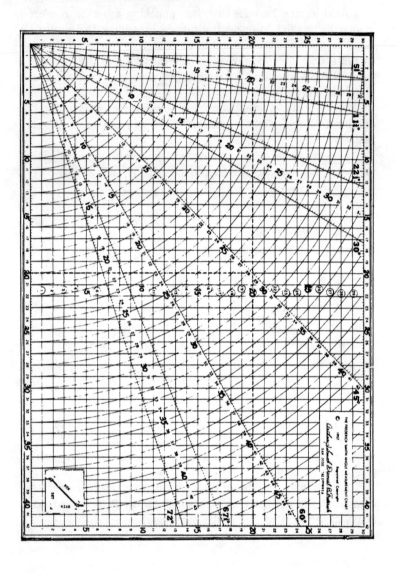

HOW TO READ THE ANGLE MEASUREMENT CHART

When looking at the chart, you will notice eight angles common to the plumbing trade. Each angle starts from a common point and extends as far as space will permit. (Each angle may be considered a miniature offset). Top and bottom numbers represent the known distance or offset that is measured on the job. The numbers on the sides of the chart represent the rise of angle. The angle run is given by the circular line passing through the angle.

To Find Run

Step 1. Assume that 20 inches is a know measurement from the center of a pipe to a point past a window and 45 degree offset fittings will be used.

Step 2. Locate on the bottom of the chart the number 20.

Step 3. Follow the dotted line up to where it intersects the 45-degree angle line. This intersection is just ¼ inch past the circular line 28, so the center to center measurement is 28¼ inches.

To Find Rise

Step 1. Let us assume that 45-degree fittings will be used on a known offset of 20 inches. Notice that, in addition to the rise numbers at the side of the chart, circled rise numbers are shown near the center. Follow line 20 up to the intersection of the angle.

Step 2. At the intersection, follow the numbered line to the side of the chart and find rise line 20. A 45-degree angle will rise 20 inches in a 20-inch offset and will develop a run of 28¼ inches.

3

Simple offsets. Letters represent known measurement. See example below.

V = vertical

H = horizontal

Figure offsets only in fitting order shown.

EXAMPLE FOR 60° ANGLE.
B=20"
C=20 x 1.155
C=RUN   A=RISE   A=20 x .577
B=SET or 20"

EXAMPLE:
D=20"   72° Vertical
20x.382=Rise
20x.619=Set
20x1.236=Run
D=Break   A=Rise   C=Run   B=Set   60° Horizontal

SINGLE ANGLE OFFSETS				COMPOUND ANGLE OFFSETS			
ANGLE	RISE(Col)	SET(Col)	RUN(Col)	ANGLE	RISE(Col)	SET(Col)	RUN(Col)
45°	B x 1.00	A x 1.00	B x 1.414	67½ V / 72 H	D x .440	D x .355	D x 1.150
60°	B x .577	A x 1.732	B x 1.155	60 V / 60 H	D x .7071	D x .7071	D x 1.414
22½°	B x 2.414	A x .414	B x 2.610	60 V / 45 H	D x 1.000	D x 1.414	D x 2.00
72°	B x .325	A x 3.077	B x 1.051	60 V / 72 H	D x .619	D x .382	D x 1.236
67½°	B x .414	A x 2.414	B x 1.082	60 V / 67½ H	D x .643	D x .493	D x 1.287
30°	B x 1.732	A x .577	B x 2.00	45 V / 72 H	D x 1.111	D x .485	D x 1.572
11¼°	B x 5.027	A x .199	B x 5.126	45 V / 67½ H	D x 1.190	D x .644	D x 1.682
5½°	B x 10.16	A x .0985	B x 10.21	45 V / 60 H	D x 1.414	D x 1.000	D x 2.00
45°	C x .707	C x .707	A x 1.414	72 V / 60 H	D x .382	D x .619	D x 1.236
60°	C x .50	C x .866	A x 2.00	67½ V / 60 H	D x .493	D x .643	D x 1.287
22½°	C x .924	C x .3826	A x 1.082	72 V / 45 H	D x .485	D x 1.111	D x 1.572
72°	C x .309	C x .325	A x 3.236	67½ V / 45 H	D x .644	D x 1.190	D x 1.682
30°	C x .866	C x .50	A x 1.155	72 V / 72 H	D x .343	D x .343	D x 1.112
11¼°	C x .981	C x .195	A x 1.02	72 V / 67½ H	D x .355	D x .440	D x 1.150
5½°	C x .995	C x .0982	A x 1.005	67½ V / 67½ H	D x .455	D x .455	D x 1.189
67½°	C x .382	C x .924	A x 2.613	30 V / 67½ H	D x 2.690	D x 1.185	D x 3.107
B IS KNOWN OFFSET MEASUREMENT				D IS KNOWN BREAK MEASUREMENT			

When figuring compound angles, always use the combination of angles in the fitting order shown.

4

SIMPLE AND COMPOUND ANGLE CONSTANTS

For simple angles, "B" is the known distance that is measured on the job. Let us assume again that the offset measurement is 20 inches, and the angle will use 60 degree fittings. (Find run center to center.)

Step 1. The left side of the simple angle chart shows the angle desired. Find 60 degrees in this column.

Step 2. At the right of the simple angle chart is shown the run column. "B" × 1.155 or substituting 20 for "B" then 20 × 1.155 = 23.10 inches.

To find rise, in the rise column find the constant that multiplied by the known measurement will give the rise. If "B" = 20 inches, then 20 × .577 = 11.54 inches or 11½ inches. If A, B, or C is known, simply substitute the known measurement multiplied by the constant in the column desired for rise, set, or run.

For compound angles, the break must be found as this measurement is the one determined offset distance.

Step 1. Determine the combination of fittings that will be used.

Step 2. The left-hand side of the compound angles has a column devoted to combinations of angles. Let us assume that we use a 60-degree horizontal Y fitting and will use a 72-degree fitting on the vertical. The ninth combination down the angle column is the one that will be used. Across the chart is found rise, run, and set columns. (Use no other combination of fittings.)

Step 3. The break or "D" is determined at 20 inches. Then 20 inches × .382 will give the rise (7.64 inches). 20 inches × .619 or 12.38 inches for the set-back distance. 20 inches × 1.236 or 24.72 inches will be run.

ROLLED OFFSETS

In order to figure this offset, the unknown measurement (X) must be calculated.

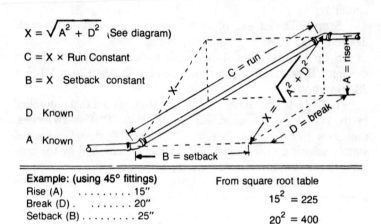

$$X = \sqrt{A^2 + D^2} \quad \text{(See diagram)}$$

C = X × Run Constant

B = X Setback constant

D Known

A Known

Example: (using 45° fittings)

Rise (A) . . . . 15″

Break (D) . . . . 20″

Setback (B) . . . . 25″

Run (C)   25 × 1.41 = 35¼″

From square root table

$15^2 = 225$

$20^2 = 400$

Add . . . 625

From page 4 find the square root of . . 625 = 25

See Page 4 for single angle constants.

Constant formulas for rolled offsets	
22½° Fittings $X = \sqrt{A^2 + D^2}$ C = × 2.61 B = X × 2.41	**45° Fittings** $X = \sqrt{A^2 + D^2}$ C = X × 1.41 B = X
60° Fittings $X = \sqrt{A^2 + D^2}$ C = X × 1.15 B = X × .58	**72° Fittings** $X = \sqrt{A^2 + D^2}$ C = X × 1.05 B = X × .325

See Page 4 for single angle constants.

EQUAL SPREAD OFFSETS

Equal spread offsets are used in parallel runs of pipe. Hangers and supports are kept in a true line a specified distance apart.

The "E" measurements must be determined. "B" is a known measurement.

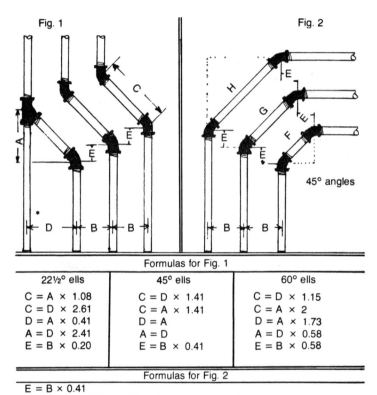

Fig. 1

Fig. 2

45° angles

	Formulas for Fig. 1	
22½° ells	45° ells	60° ells
C = A × 1.08	C = D × 1.41	C = D × 1.15
C = D × 2.61	C = A × 1.41	C = A × 2
D = A × 0.41	D = A	D = A × 1.73
A = D × 2.41	A = D	A = D × 0.58
E = B × 0.20	E = B × 0.41	E = B × 0.58

Formulas for Fig. 2

E = B × 0.41
F = G − (B × 0.41 × 2) or G = F + (B × 0.41 × 2)
G = H − (B × 0.41 ×) or H = G + (B × 0.41 × 2)

* Note: Do not confuse D with B measurement.

AIR GAPS—BACK FLOW PREVENTERS—VACUUM BREAKERS

Air Gap Fittings. These fittings are designed for an air break in waste lines. The air break is fixed in the fitting and is tested and approved by nationally recognized testing agencies for back flow and spillage.

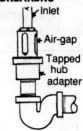

Air gap fittings are designed for free flow. (Do not improvise; ask for "research approval.")

Vacuum Breakers. Vacuum breakers are designed to prevent a possible contamination of a water supply. The construction depends on an inner check that allows the water to flow through, but does not permit a siphon action as the top is open to the atmosphere. Used between the inlet to a plumbing fixture and the water supply. Install 6 inches above overflow rim of fixture (see conditions).

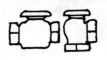

Pressure Type Vacuum Breakers. These fittings are designed to isolate an entire system and will work under pressure. Install not less than 12 inches above ground or floor.

Check with your local inspection department for conditions and requirements when installing such equipment. Industrial lines must be separated from potable water supply lines.

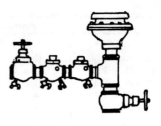

Back Flow Preventers. These may be double-check valve assemblies and are used on low pressure boilers, steam cookers, food or beverage dispensers, or any place that would allow a possible entry of non-toxic materials or liquid into the water supply system. Install not less than 12 inches above point of usage. Check conditions of installation for different types of use.

AIR GAP FITTINGS

Air gaps are required where there is any possibility of contamination of food or water. Check for research approval.

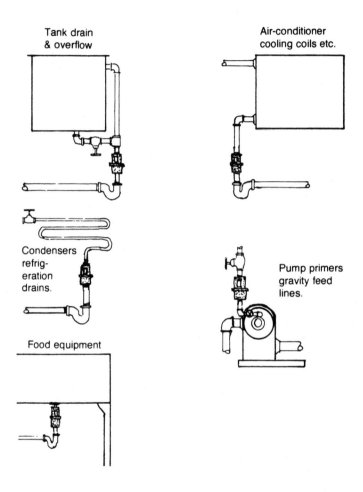

Tank drain & overflow

Air-conditioner cooling coils etc.

Condensers refrigeration drains.

Pump primers gravity feed lines.

Food equipment

Approved free flow air gap fittings are designed for installations that require a physical separation from fixture or appliance outlet. To entry in sanitary sewer.

Air gap fittings are designed for connecting wastes to drains that require a separation between waste and trap, such as special wastes and indirect connections.

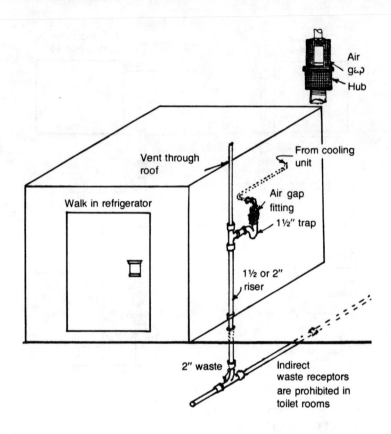

Air gap fittings may be used only when fixture wastes are above floor. Do not install in concealed places, below floor level, or in rooms that have no steady occupancy.

Some uses are for water cooling wastes, steam tables, condensers, tank drains, and ice boxes.

ASPIRATOR UNIT

A suction device used in hospitals and similar institutions. Check local codes and health requirements for installation.

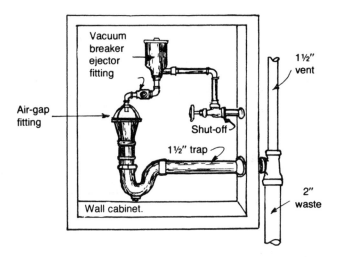

All hospital equipment must be protected by approved vacuum breakers and air gaps. Install according to code and manufacturer's specifications.

Check code requirements for location of all medical, mortuary, surgical, or similar installations.

Where vacuum breakers and air gaps are part of the equipment, check for approval of the unit, location of vacuum breakers, and ascertain that it will not be subject to back flow or back pressure at any time.

FORMULAS FOR AREAS

Square
$A = b \times h$

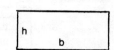

Rectangle
$A = b \times h$

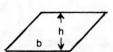

Rhombus
$A = b \times h$ An equilateral parallelogram
with oblique angles.

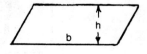

Rhomboid
$A = b \times h$ A parallelogram with
oblique angles and
only opposite sides
equal.

Trapezoid
$$A = \frac{h \times (a + b)}{2}$$

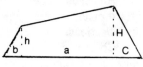

Trapezium
$$A = \frac{(H + h)a + bh + CH}{2}$$

Right angle triangle
$$A = \frac{b h}{2}$$

Isosceles triangle
$$A = \frac{bh}{2}$$
A triangle with two equal sides.

Area of an ellipse

$$A = \pi\,ab, \quad \text{where } a = \frac{D}{2}$$

$$\text{and} \quad b = \frac{d}{2}$$

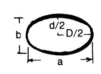

Formulas for volume of solids $A = \pi\,Dh$
Area of the surface of cylinder
Cubical contents or volume of cylinder

$$V = .7854 \times D^2\,h$$
$$V = \pi\,R^2\,h$$

Cubical contents or volume of cube or rectangular solid.

$$V = Whl$$

Area of surface of pyramid
n = number of sides
b = length of one side
s = slant height

$$A = \frac{nbs}{2}$$

Area of surface of cone $A = \dfrac{\pi Ds}{2}$

Volume of pyramid or cone $V = \dfrac{bh}{3}$

Area of surface frustrum of cone

$$A = \frac{\pi\,s\,(D+d)}{2}$$

Volume of frustrum of cone

$$V = \frac{\pi\,h\,(D^2 + d^2 + Dd)}{12}$$

FORMULAS FOR THE CIRCLE

Where C = circumference

Where R = radius

Where D = diameter

Where A = area.

The symbol π = 3.1416

Circumference = D π

Circumference = 2 π R

Radius = $\dfrac{D}{2}$

Radius = $\dfrac{C}{2\pi}$

Area = π R^2

Area = $\dfrac{D^2 \pi}{4}$

Area = $\dfrac{CR}{2}$

Area = D^2 • .7854

Diameter = $\dfrac{C}{\pi}$

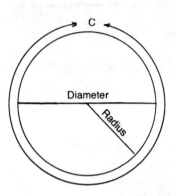

The formula most used to find area is the diameter squared multiplied by the decimal .7854.

A common method to find circumference is to multiply the diameter by 3 1/7.

Radius equals ½ the diameter.

TO FIND AREA OF REGULAR POLYGON

A regular polygon is a figure with any number of sides, all equal to each other in length.

Name of Polygon	Number of sides	K Constant for area
Triangle	3	.433
Square	4	1.000
Pentagon	5	1.720
Hexagon	6	2.598
Heptagon	7	3.364
Octagon	8	4.428
Nonagon	9	6.182
Decagon	10	7.366
Undecagon	11	9.366
Dodecagon	12	11.196

Formula: $A = S^2 K$

Where; s = length of one side
k = given constant

Square one side and multiply by the constant.

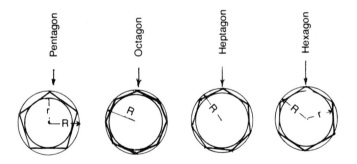

FRACTIONS AND DECIMAL EQUIVALENTS

Fractions and Decimal Equivalents			
1/64	.015625	33/64	.515625
1/32	.03125	17/32	.53125
3/64	.046875	35/64	.546875
1/16	.0625	9/16	.5625
5/64	.078125	37/64	.578125
3/32	.09375	19/32	.59375
7/64	.109375	39/64	.609375
1/8	.125	5/8	.625
9/64	.140625	41/64	.640625
5/32	.15625	21/32	.65625
11/64	.171875	43/64	.671875
3/16	.1875	11/16	.6875
13/64	.203125	45/64	.703125
7/32	.21875	23/32	.71875
15/64	.234375	47/64	.734375
1/4	.25	3/4	.75
17/64	.265625	49/64	.765625
9/32	.28125	25/32	.78125
19/64	.296875	51/64	.796875
5/16	.3125	13/16	.8125
21/64	.328125	53/64	.828125
11/32	.34375	27/32	.84375
23/64	.359375	55/64	.859375
3/8	.375	7/8	.875
25/64	.390625	57/64	.890625
13/32	.40625	29/32	.90625
27/64	.421875	59/64	.921875
7/16	.4375	15/16	.9375
29/64	.453125	61/64	.953125
15/32	.46875	31/32	96875
31/64	.484375	63/64	.984375
1/2	.5	1	1.00

1% pitch equals ⅛"
per foot roughly.
Actually, 1% will
rise 1' in each 100!

2% pitch equals ¼"
per foot roughly.
Actually 2% will rise
2' in each 100 ft.

BAPITISTRIES, ORNAMENTAL POOLS, FOUNTAINS, AND SIMILAR CONSTRUCTION

Consult local inspection department regulations.

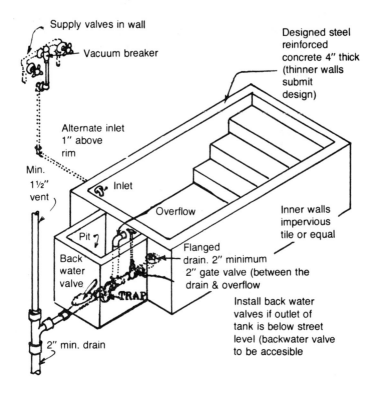

All built-on-the-job construction must have a suitable waste and overflow, so constructed and installed to maintain a leakproof joint. Where cast into concrete, flanged drains and overflows are required. See next page.

BUILT-ON-THE-JOB FIXTURES

All Roman baths, baptistries, aquariums, or similar construction must be designed properly to make a sanitary leak-proof fixture and be adequately supported on its own foundation.

Concrete must be designed for 3000 psi compressive strength, with approved waterproof admixture. Steel reinforcing must be not less than ⅜ inches re-bars 6 inches mesh, imbedding of steel in center of 4 inch wall. Walls of concrete to be plastered with a cement mortar ¾ inch thick with approved admixture. Tile or equal surface on above mix.

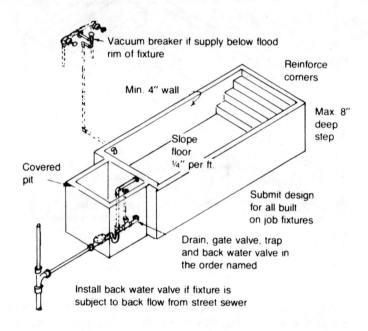

Steel may be used instead of concrete when designed. Weld all joints, spot weld metal lath to steel wall. Red lead interior and metal lath before applying mortar. Bolt and weld or braze all wastes and overflows. See details next page.

ROMAN BATH DETAILS

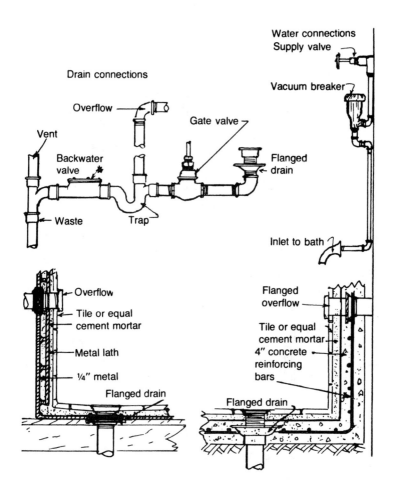

* Install back water valve if drain is below level of top of street.

Support all built-on-the-job receptacles on their own foundations. Fill with water and test for 24 hours before approval.

BAR SINK AND COOLING COUNTER INDIRECT WASTE HOOKUPS

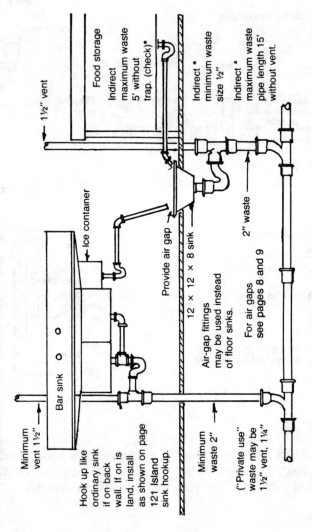

Minimum vent 1½"

Hook up like ordinary sink if on back wall. If on island, install as shown on page 121 Island sink hookup.

Bar sink

Ice container

1½" vent

Food storage

Indirect maximum waste 5' without trap. (check)*

Provide air gap

12 × 12 × 8 sink

Air-gap fittings may be used instead of floor sinks.

For air gaps see pages 8 and 9

Indirect * minimum waste size ½"

Indirect * maximum waste pipe length 15' without vent.

2" waste

Minimum waste 2"

("Private use" waste may be 1½" vent, 1¼".)

Locate all floor sinks as close as possible to drain outlet of the fixture requiring special wastes. * Check local regulations.

20

BATHTUB INSTALLATION

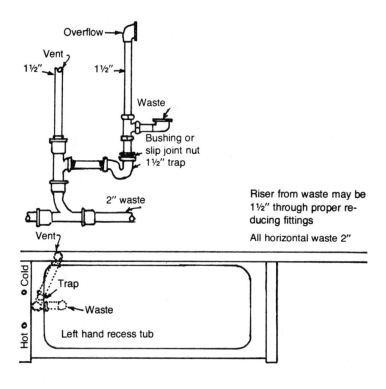

Overflow

Vent
1½"

1½"

Waste

Bushing or
slip joint nut
1½" trap

2" waste

Riser from waste may be
1½" through proper re-
ducing fittings

All horizontal waste 2"

Vent

Cold

Trap

Waste

Hot

Left hand recess tub

Rough waste and vent off to one side to allow for trap. Trap should be at least 6 inches below the finish floor to allow for waste and overflow.

Water pipe to tub filler should center over overflow, spout 6 inches above rim of tub.

Provide access panel on concrete floors or on two-story work.

Check rough-in measurements on all tubs.

ROMAN BATHS (MANUFACTURED TYPE)

Bathtubs and one-piece Roman baths have been a recent development of the fiberglass industry. Check commercial standards for fiberglass.

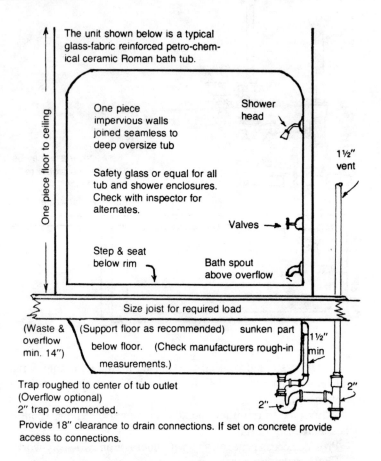

The unit shown below is a typical glass-fabric reinforced petro-chemical ceramic Roman bath tub.

One piece floor to ceiling

One piece impervious walls joined seamless to deep oversize tub

Safety glass or equal for all tub and shower enclosures. Check with inspector for alternates.

Step & seat below rim

Shower head

1½" vent

Valves →

Bath spout above overflow

Size joist for required load

(Waste & overflow min. 14")

(Support floor as recommended) below floor. (Check manufacturers rough-in measurements.)

sunken part

1½" min

Trap roughed to center of tub outlet
(Overflow optional)
2" trap recommended.

2"

2"

Provide 18" clearance to drain connections. If set on concrete provide access to connections.

See commercial standard 221-59 for bathtubs. Check manufacturer's rough and finish specifications for fiberglass installation procedure.

BATHTUB INSTALLATION

1. Support. Install proper blocking for leveling tub.

2. Waste and overflow. Install 1½-inch waste and overflow to 1½-inch trap.

3. Water Supplies. Install tub valves and filler above overflow rim of tub. Minimum of tub filler 1 inch above rim of tub. Tub supplies normally rough in 8 to 12 inches above rim. Spout of filler usually from 3 to 6 inches above rim.

4. Concrete floors and concealed work. Install access door to slip joint tub connections. Minimum size trap door 12 by 12 inches. Where tubs are back to back, install doors on outside walls and in position for access to connections.

5. Trap arms and traps. Galvanized pipe and durham traps may be used if 6 inches or more above the ground. When below ground level install brass, cast iron or copper traps as specified by code. No galvanized pipe allowed below ground for drain pipe unless allowed by ordinance.

6. Traps under concrete floors. Install block out in floor to receive waste and overflow connection. Block out should be 8 inches wide and 12 inches long and run to partition line. On access doors to an outside wall, run 12 inches wide along partition line to outside wall.

7. Built up tubs. Built-on-the-job tubs are prohibited unless for special conditions such as baptistries.

8. Special tubs. When special conditions are encountered, consult with the local inspection department for requirements. Submit detailed plans and specifications for design and construction. Install back water valves on all plumbing fixtures that are below street level.

9. Water supplies for special tubs. All water fountains, fish ponds, wading pools or similar installations must be provided with approved air gap installed 1 inch above extreme overflow rim of the fixture and 6 inches above the ground.

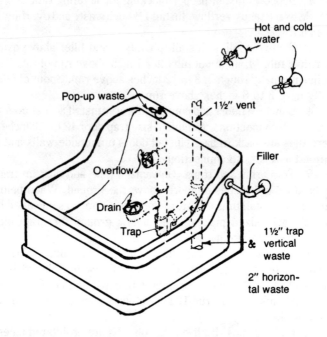

Sitz baths are commonly used in hospitals and similar places where it is not practical to immerse the whole body.

Where water inlets may allow a cross connection, vacuum breakers must be installed. Install an access panel to drain hookup and slip joint connections.

Rough in for a sitz is similar to an ordinary bathtub. Follow manufacturer's measurements and specifications. Check local codes and health requirements.

FIXTURES BELOW STREET LEVEL

All plumbing installed so that public sewers, in any possible way, could fill or overflow a plumbing fixture in premises requiring inspection requires back water valves.

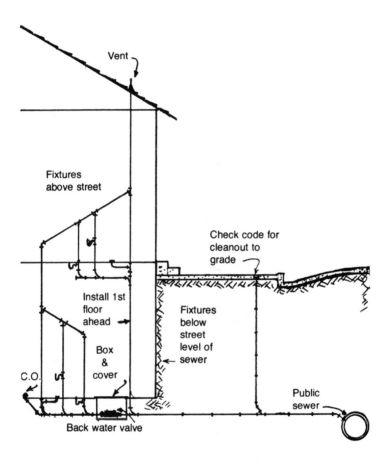

Install back-water valve to serve only those fixtures subject to back flow from public sewer.

For a back-water valve to be accessible for inspection and repair, install a box and cover if below grade.

BOILER BLOW-OFF PIPING

Rule 1. Relief pipe one size lager than largest inlet, extend full size above roof.

Rule 2. Outlet one size larger than largest inlet.

Rule 3. Blow-off tank must be of sufficient size to cool contents to 140 degrees or less.

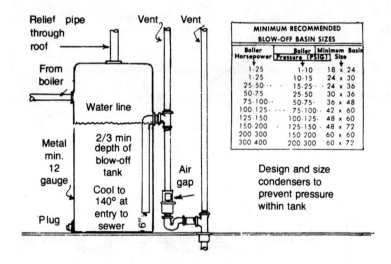

MINIMUM RECOMMENDED BLOW-OFF BASIN SIZES		
Boiler Horsepower	Boiler Pressure (PSIG)	Minimum Basin Size
1-25	1-10	18 x 24
1-25	10-15	24 x 30
25-50	15-25	24 x 36
50-75	25-50	30 x 36
75-100	50-75	36 x 48
100-125	75-100	42 x 60
125-150	100-125	48 x 60
150-200	125-150	48 x 72
200-300	150-200	60 x 60
300-400	200-300	60 x 72

Design and size condensers to prevent pressure within tank

Keep all inlets above the water line.

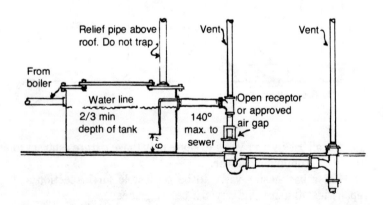

Note: Minimum 12-gauge iron, protected from external corrosion, may be used in construction of condensers or reinforced engineered concrete.

BED PAN STEAMER

Check manufacturer's measurements and specifications for wall or floor connections.

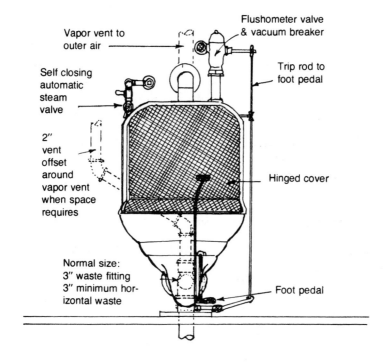

Vapor vent to outer air

Flushometer valve & vacuum breaker

Self closing automatic steam valve

Trip rod to foot pedal

2" vent offset around vapor vent when space requires

Hinged cover

Normal size: 3" waste fitting 3" minimum horizontal waste

Foot pedal

All water inlets must be protected by vacuum breakers.

Atmospheric or vapor vents are continued to the outer air separate. No connection is permitted to a fixture vent.

Check local codes and health regulations for installation requirements.

Waste water not over 140 degrees to drainage system.

BIDET INSTALLATION

Rough in waste below floor similar to bathtub. Connections are exposed. Tailpiece is usually 1¼ inches. Water supply not less than ½ inches, in wall.

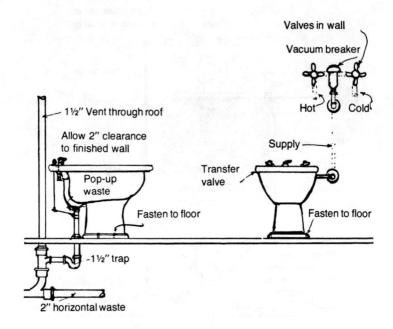

When installing or roughing in fixtures, always follow manufacturer's measurements and specifications.

Water pipe may rough in directly behind bidet or may rough in off center as shown.

In all cases, install vacuum breaker. Hot- and cold-water valves are installed in the wall. They must be securely strapped to prevent moving. Grout to floor after setting and testing.

BURIED PIPING

Cast Iron. Cast iron may be installed anywhere that drain pipe is required. Protection is required where specified by job specifications. Do not imbed in concrete walls or footings.

Vitrified Clay. Do not install vitrified clay closer than 2 feet to a building. Allow for at least 12 inches coverage above top of bell. Install clay pipe so that barrel of pipe rests on bottom of trench; dig out for all bell-holes. Clay pipe is not recommended where rocky soil is encountered. If clay pipe is installed where soil conditions are not suitable, set in sand or other approved aggregate. Fill to 12 inches above bell. Do not use mechanical equipment until backfilled 12 inches by hand.

Bituminus Fiber Pipe. Install fiber pipe with approved manufactured fittings. Allow at least 12 inches fill above pipe. Manufacturer's specification installation requirements should be followed.

Cement Asbestos. Allow 12 inches coverage over pipe. Install as to manufacturer's specifications for soil conditions. Use only approved sewer pipe and fittings as required by code.

Clay and Water In Same Trench. When installing water pipe in same trench, keep water above clay pipe at least 12 inches away from side of trench on separate shelf.

Water Pipe. Protect as required by job specification. Materials allowed by code.

Gas Pipe. Protect as specified by code. Do not install gas piping under building slab. Materials: brass, copper, wrought iron, steel. No copper tubing allowed for gas pipe, except when allowed as tubing connectors.

MINIMUM COVERAGE FOR BURIED PIPE ACCORDING TO MATERIAL

Cast iron not shown as no minimum coverage is required. Protect to job specification.

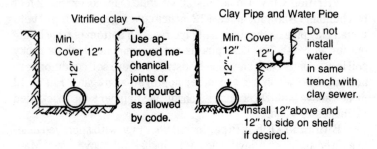

Keep all steel drains 6 inches above ground level. Keep all DWV copper 6 inches above ground level.

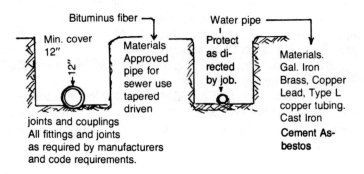

Keep steel and DWV copper drains above ground.

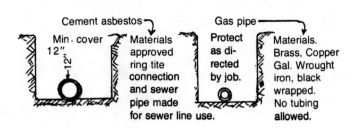

BYPASS IN DRAINAGE SYSTEM

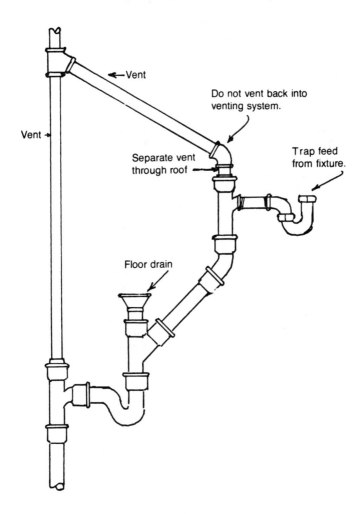

Under certain conditions, this installation will permit escape into the building. To correct, run vent from fixture trap separate. Any plumbing installation that will allow this condition is prohibited.

CORNER LOTS, INTERIOR LOTS:
BUILDING (HOUSE) SEWERS AND MAIN VENTS

Each building shall have a separate connection to the street within the exception noted below.

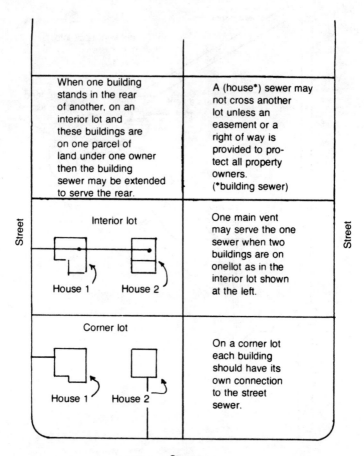

When one building stands in the rear of another, on an interior lot and these buildings are on one parcel of land under one owner then the building sewer may be extended to serve the rear.	A (house*) sewer may not cross another lot unless an easement or a right of way is provided to protect all property owners. (*building sewer)
Interior lot — House 1, House 2	One main vent may serve the one sewer when two buildings are on onellot as in the interior lot shown at the left.
Corner lot — House 1, House 2	On a corner lot each building should have its own connection to the street sewer.

Street

Street

Street

Street

CLEANOUT REQUIREMENTS

An improperly installed cleanout has no value. A good design avoids unnecessary cleanouts.

1. Access. Provide access within 20 feet of any cleanout or extend to outside. Minimum crawl space should have 18 inches vertical clearance. No cleanout should be located in a space less than 30 inches in width (allow room enough to work).

2. Inaccessible. Any cleanout is considered inaccessible when located under concrete, too close to walls or foundations, or located in any space that does not have adequate clearances.

3. Main Drains. Provide 18 inches clearance in front and vertically. Allow 12 inches clearance in front for 2 inches or smaller waste-line cleanouts.

4. Where Required. At base of stacks, upper end of horizontal building drains or its branches, at changes of direction specified by code, at junctions of building drains and house sewers and at property lines when required by code.

5. Intervals. Maximum distance apart should be not more than 100 feet measured by the developed run of pipe, including the cleanout extension. On some installations such as a combination waste and vent system, cleanouts may be required on intervals of 50 feet due to its low scouring potential.

6. Size of Cleanouts. Cleanout must be at least as large as the line it serves. A 4-inch cleanout will normally serve any larger line, unless specifically required. Check local regulations.

7. Wall or Floor Cleanouts. Provide adequate working space, maintain clearance required by code. Check for approved types of cleanouts, access boxes, or wall plates. Countersunk plugs may be used for flush installations when specified.

8. Two-Way Cleanouts. May be substituted for upper terminal cleanouts when approved by the Administrative Authority. See the next page.

9. Prohibited Locations. No cleanouts permitted inside sumps or interceptors.

10. Approvals. Check for material, size, and type.

CLEANOUTS

Maximum spacing of cleanouts must not exceed 100 feet. Install cleanouts at the base of all stacks, terminations of all horizontal runs and all changes of direction required by code.

When you are unable to maintain 18 inches clearance extend to an accessible spot, maintain 18 inches for all main cleanouts.

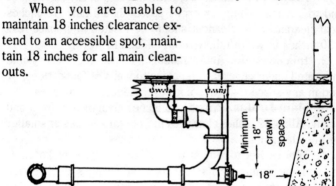

Install C.O. at property line where required. An approved two way cleanout may be used within 2 feet of house instead of upper terminal cleanout.

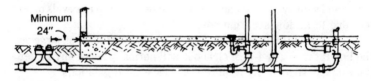

If code definition of vertical is 45 degrees, 60 degrees, or 72 degrees, no cleanout is required on piping installed in vertical position.

In concealed places above one story, cleanouts may be omitted. See code.

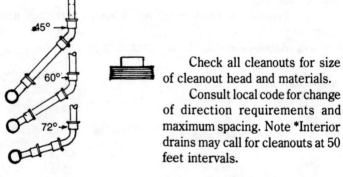

Check all cleanouts for size of cleanout head and materials.

Consult local code for change of direction requirements and maximum spacing. Note *Interior drains may call for cleanouts at 50 feet intervals.

Cleanouts are required on all upper terminals of the horizontal building drain and its branches, on changes of direction greater than 22½ degrees, and at intervals of 100 degrees or fraction thereof. Check local regulations.

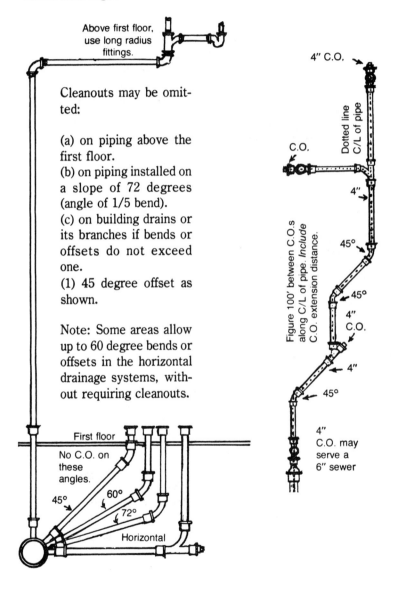

Above first floor, use long radius fittings.

4″ C.O.

C.O.

Dotted line C/L of pipe

Figure 100′ between C.O.s along C/L of pipe. Include C.O. extension distance.

4″

45°

45°

4″ C.O.

4″

45°

4″ C.O. may serve a 6″ sewer

Cleanouts may be omitted:

(a) on piping above the first floor.

(b) on piping installed on a slope of 72 degrees (angle of 1/5 bend).

(c) on building drains or its branches if bends or offsets do not exceed one.

(1) 45 degree offset as shown.

Note: Some areas allow up to 60 degree bends or offsets in the horizontal drainage systems, without requiring cleanouts.

First floor

No C.O. on these angles.

45°   60°   72°

Horizontal

Cleanouts required if less than 72 degrees (angle 1/5 bend).

CLOTHESWASHERS, SELF-SERVICE

Note example allows 4 units per clotheswasher. Use "approved" divisional pattern fittings when installing clotheswashers back to back.

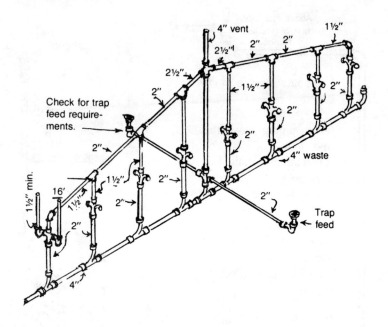

Clotheswasher standpipes length 18 inches minimum, 30 inches maximum. Vent intersections minimum 6 inches above overflow level of top of standpipe receptors.

Note: Self-service laundries have been a subject of much criticism because of inadequate design. The first rule is to obtain exact waste GPM discharge and design accordingly. In the event this is not known, assigning 4 units for each clotheswasher roughed in would be ample for most clotheswashers in this category.

1. Check for actual discharge in GPM if possible.

2. Check appliance "approval" for cross-connection control.

3. Design water supply for public-use demand.

4. Check for gas dryers, find total Btu and size according to demand and length of run.

5. Provide for ample combustion air into room. Check gas vents for size, material, and clearance.

CLOTHESWASHER INSTALLATION

1. Roughing In. Install 2 inches horizontal waste. Vertical waste riser may be 1½ inches. Minimum trap size 1½ inches. Minimum vent size 1½ inches. Rough waste height 6 to 12 inches from finished floor. Install hot and cold water in wall behind washer. Recommended height of valves in wall 48 inches from finished floor, 8 inches apart, center to center, with the cold water on right of center and hot to the left. Run water from valves down and behind into recess provided for the hoses. Strap and secure all piping.

2. Recess. Frame in minimum recess in wall for piping 14 inches wide and 30 inches high. Check height.

3. Supports. Provide supports as required by the manufacturer.

4. Standpipe. Install standpipe not less than 18 inches or more than 30 inches long. Do not locate trap for clotheswasher under the floor.

5. Water Supplies. All washers must have water inlets into washer protected by approved air gaps as required by code.

6. Each washer must have a separate trap, except that if alongside a tray it may spill into tray.

7. Finish. Level and secure according to manufacturer's specifications.

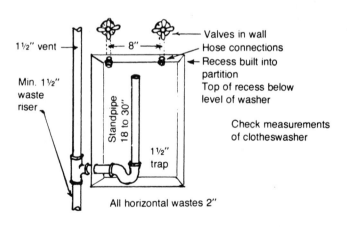

1½" vent

Min. 1½" waste riser

Standpipe 18 to 30"

8"

Valves in wall
Hose connections
Recess built into partition
Top of recess below level of washer

Check measurements of clotheswasher

1½" trap

All horizontal wastes 2"

COMBINATION WASTE AND VENT SYSTEM

Note: Cleanouts are required on any branch of trap arm serving more than a single trap.

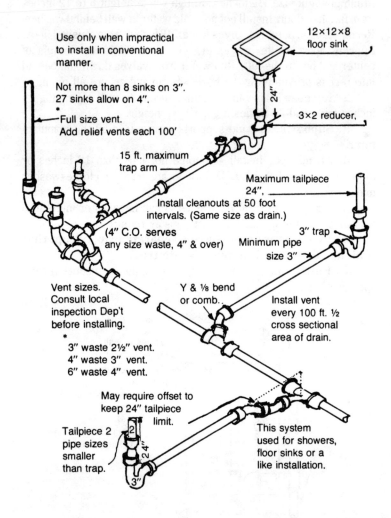

Use only when impractical to install in conventional manner.

Not more than 8 sinks on 3". 27 sinks allow on 4".
*
Full size vent. Add relief vents each 100'

12×12×8 floor sink

24"

3×2 reducer,

15 ft. maximum trap arm

Maximum tailpiece 24",

Install cleanouts at 50 foot intervals. (Same size as drain.)

(4" C.O. serves any size waste, 4" & over)

3" trap

Minimum pipe size 3"

Vent sizes. Consult local inspection Dep't before installing.
*
3" waste 2½" vent.
4" waste 3" vent.
6" waste 4" vent.

Y & ⅛ bend or comb.

Install vent every 100 ft. ½ cross sectional area of drain.

May require offset to keep 24" tailpiece limit.

Tailpiece 2 pipe sizes smaller than trap.

2"

24"

3"

This system used for showers, floor sinks or a like installation.

If a cleanout is installed on a vent to serve waste, install same size as for drains.

CONDENSATE DRAINAGE TO SANITARY SEWER

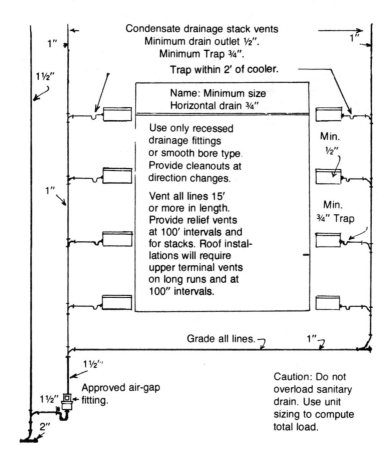

Condensate drainage stack vents
Minimum drain outlet ½".
Minimum Trap ¾".

Trap within 2' of cooler.

Name: Minimum size
Horizontal drain ¾"

Use only recessed
drainage fittings
or smooth bore type.
Provide cleanouts at
direction changes.

Vent all lines 15'
or more in length.
Provide relief vents
at 100' intervals and
for stacks. Roof instal-
lations will require
upper terminal vents
on long runs and at
100" intervals.

Min.
½"

Min.
¾" Trap

Grade all lines.

1"

1"

1"

1"

1½"

1"

1½"

Approved air-gap
fitting.

1½"

2"

Caution: Do not
overload sanitary
drain. Use unit
sizing to compute
total load.

Condensate drainage varies with climate and sections of the country. See the next page for a safe table that will prove satisfactory anywhere. Pipe size equals total area of unit drain outlets.

CONDENSATE DRAIN TABLE

Definition of condensate waste: waste water caused by the formation of moisture on the outside of coils or other surfaces.

Drain Outlet Size	Total Connections Shown Below Main Waste Size								Horizontal Main Waste	Fixture Assigned Unit Value
	3"	2½"	2"	1½"	1¼"	1"	¾"	½"•		
½"	36	25	20	9	6	4	2	0	½"•	¼
¾"	16	11	7	4	3	1	1	0	¾"	½
1"	9	6	4	2	1	1	0	0	1"	¾
1¼"	6	4	2	1	1	0	0	0	1¼"	1
1½"	4	2	1	1	0	0	0	0	1½"	2
2"	2	1	1	0	0	0	0	0	2"	5
2½"	1	1	0	0	0	0	0	0	2½"	9
3"	1	0	0	0	0	0	0	0	3"	14

Smaller drains may be added not to exceed main horizontal cross-sectional discharge area.

• No ½" Horizontal

How to read the table. Example (using ½-inch drain outlet size the assigning ¼ unit valve for each condensate drain) see the preceding page showing eight coolers.

1. Main waste size for eight coolers will be 1½ inches. See allowable ½ inch drains under main waste size.

2. Vertical wastes. 4 drain connections will be 1 inch minimum for each drainage stack.

3. Drainage stack vents will be equal in size to drainage stack size or 1 inch minimum.

4. Intersection of horizontal to vertical will be increased to 1½ inches. See typical sizing.

5. Allow eight times ¼ or 2 units for waste loading value at house drain entry.

All drainage pipe and fittings should be smooth bore such as Durham or copper. All changes of directions should be supplied with cleanouts as directed by code requirements governing other wastes. Do not use ½-inch pipe for drains unless for single vertical drain under 15 feet in length.

Minimum horizontal waste and trap should be not less than ¾ inches in size. Traps may be bent soft temper copper with a minimum seal of 2 inches. Vent all risers or drops 15 feet or more in length.

TYPICAL CROSS CONNECTIONS

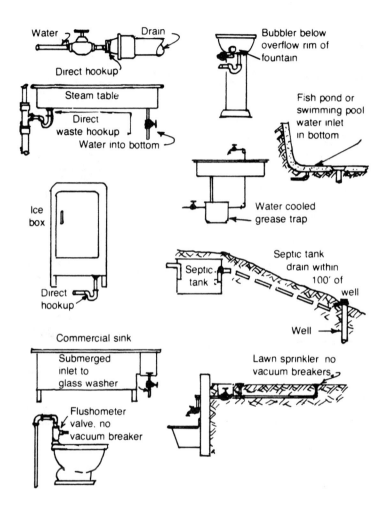

Any possible connection between a plumbing fixture and the domestic water supply is a cross connection. Do not connect industrial water to domestic water without adequate back flow prevention devices.

CROSS CONNECTIONS

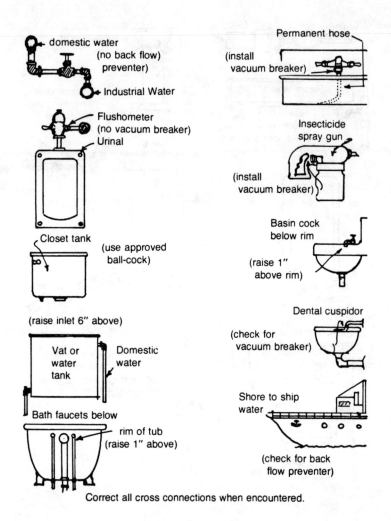

domestic water
(no back flow)
preventer)

Industrial Water

Flushometer
(no vacuum breaker)
Urinal

Closet tank
(use approved
ball-cock)

(raise inlet 6" above)

Vat or water tank

Domestic water

Bath faucets below

rim of tub
(raise 1" above)

Permanent hose

(install vacuum breaker)

Insecticide spray gun

(install vacuum breaker)

Basin cock below rim

(raise 1" above rim)

Dental cuspidor

(check for vacuum breaker)

Shore to ship water

(check for back flow preventer)

Correct all cross connections when encountered.

42

DEFINITIONS OF THE PLUMBING SYSTEM

1. Stack vent. Main vent through roof drainage stack.

2. Back vent. Fixture vent connecting to a vent stack or drain stack.

3. Drain or soil stack. Vertical drain extending more than one story.

4. Waste branch. Horizontal line off building drain.

5. Building drain. Receives all wastes within the building and conveys them to the house sewer.

6. House sewer cleanout. Serves the house sewer.

7. *House sewer from the lateral to a point 2 feet from the building or foundation line.

8. Test Y and C.O. . . A. "Y"-fitting installed for test purpose and later extended for a lateral cleanout.

9. Lateral. The sewer to the property line from the public sewer.

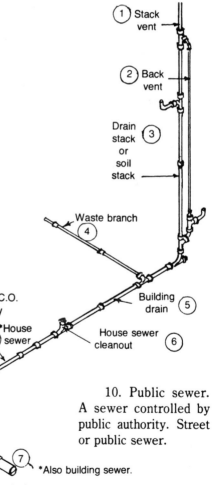

10. Public sewer. A sewer controlled by public authority. Street or public sewer.

See vent and waste stacks.

Air Gap. An air gap as applied to plumbing is a vertical physical separation between a water supply and a sewer drain, receptacle, plumbing fixture or other device that in any way could possibly contaminate a domestic water supply. An air gap is also a separation between a waste from a food container and its connection to the sewer.

Approved. Approved means accepted by the administrative authority for a proposed use.

Backflow. Backflow is the entry of other than potable water into a domestic water supply.

Backflow Connection. Backflow connection is the connection that would permit the entry into another piping system of liquids not intended for such systems.

Backflow Preventer. A device that prevents backflow.

Back-siphonage. Back-siphonage is the action of backflow due to a negative pressure or elevation difference that would allow a source of contamination to enter a potable water supply.

Boiler Blow-Off. A boiler blow-off is the drain from the boiler that allows the emptying of the contents of the boiler.

Branch. A branch is that part of the piping system other than a stack, riser or main.

Branch Vent. A branch vent is a vent or vents connecting with a vent stack.

Building. A building is a structure designed and erected for the housing, shelter, enclosure, and support of persons, animals, or property.

Building Drain. The building drain is the main horizontal drain within the building collecting all soil and wastes and conveying all such to a point 2 feet from the building wall.

Building Sewer. Building sewer is the house connecting to the building drain starting 2 feet from the house and conveying soil and wastes to the point of disposal, whether public sewer, private sewer, individual disposal system or any other means of disposal.

Cesspool. A cesspool is a lined pit or excavation in the ground built to retain organic matter and solids, but allowing the liquids to seep into the ground through the bottom and sides.

Code. A code is the law regulating the plumbing installations when used in reference to the rules and regulations legally adopted by agencies that have such authority.

Combination Waste and Vent System. A combination waste and vent system is a horizontal drain pipe in which the upper half of the drain acts as the vent. (see illustration)

Common. Serving more than one building, fixture, appliance, or system.

Continuous Vent. A vent continuing upward from the drain to which it connects.

Continuous Waste. A continuous waste is a drain connecting compartments of a combination fixture to its trap or, when permitted, allows more than one fixture connection to a common trap.

Cross Connection. A cross connection is any arrangement whereby unclean, polluted, contaminated liquid or matter may enter a potable water supply in any way.

Dead End. A dead end is a pipe 2 feet or more in length branching from a soil, waste, or vent pipe terminating in a plug or other closed fitting.

Developed length. The developed length of a pipe is its length along the center line of the pipe and fittings.

Diameter. Diameter as applied to pipe is the nominal diameter specified by the manufacturer.

Domestic Sewage. Domestic sewage is ordinary wastes from living quarters not requiring special treatment before entering the public sewer or private disposal system.

Drain. A drain is any pipe that carries wastes or waterborne wastes in a building drainage system.

Drainage System. A drainage system includes all pipes within public or private premises installed to convey wastes from the plumbing system to a legal point of disposal, but does not include the public sewer system or its treatment plant.

Durham System. A durham system is a threaded pipe and fitting installation using a recessed drainage fitting approved for that type of pipe and providing a smooth interior waterway.

Effective Opening. The effective opening is the minimum cross-sectional area of the object at the point of discharge. (Diameter of circle; or if not circular the diameter of a circle of equivalent cross-sectional area.)

Existing Work. Existing work is that part of a plumbing system that has been installed prior to the effective data of the current code.

Fixture Branch. A fixture branch is a section of pipe serving several fixtures.

Fixture Drain. A fixture drain is a drain connection from a fixture trap to another drain.

Fixture Supply. A fixture supply is the water outlet installed to serve the fixture.

Fixture Unit. A fixture unit is a number assigned to a water or waste pipe expressing the volume rate of flow or discharge capacity of such pipes, including the load allowance for vent pipes.

Fixture Unit Flow Rate. Fixture unit rate of flow is the total discharge in gpm of a single fixture, divided by 7.5, provides the flow rate of that particular fixture as a unit of flow. Fixtures are rated in multiples of this unit of flow.

Flood Level Rim. The flood level rim is the extreme top edge or rim of the object that when filled with water will overflow.

Flooded. Flooded means that a fixture is full to the point of rim overflow.

Flush Valves. A flush valve is a device located in the bottom of a tank for the purpose of flushing the intended fixture.

Flushometer Valve. A flushometer valve is a valve designed to discharge a predetermined amount of water to a fixture for flushing purposes and is actuated by water pressure.

Grade. Grade as related to the plumbing system is the amount of fall or pitch per foot required to properly drain a horizontal line.

Horizontal Branch. A horizontal branch is a drain pipe extending horizontally from a waste or soil stack with or without vertical sections or branches and conveys wastes into the stack or building drain.

Horizontal Pipe. A horizontal pipe is any position or angle not considered or defined as vertical in the installation of piping.

House Drain. *See* building drain.

House Sewer. *See* building sewer.

Indirect Waste Pipe. An indirect waste pipe is a pipe that does not connect directly to the drainage system, its waste discharging into a receptacle and separated from the receptacle of the direct drainage system by a physical separation or an approved air-gap fitting.

Individual Vent. An individual vent is a vent pipe serving a fixture trap, connecting to other vent piping or terminating separately to the outer air.

Industrial Waste. Industrial wastes means waste from industrial or commercial processes. (it is not domestic waste.)

Insanitary. Insanitary as applied to the plumbing system means faulty plumbing, nonscouring, insanitary plumbing fixtures, cross connections, traps not protected against siphonage, back-pressure or evaporation, and any or all conditions that would be detrimental to health.

Interceptor. An interceptor is a device designed to separate and retain matter from normal wastes and still permit normal sewage to enter the drainage system by gravity.

Liquid Wastes. Liquid wastes are wastes that are discharged from a plumbing system that does not receive fecal matter.

Lot. A lot is a lawfully recorded parcel of land in possesion of a legal owner upon which a building is erected or constructed and is the site on which work is performed.

Main. A main is the principal pipe to which all branches are connected.

Main Vent. A main vent is the principal vent pipe of the venting system to which branch vents may be connected.

Main Sewer. Main sewer is a public sewer or common sewer controlled by public authority.

Offset. Offset as defined in plumbing means an arrangement of fittings in a line of pipe that will bring the line over to a predetermined distance and in a parallel line.

Plumbing. Plumbing is the business, trade or work within the scope of work performed by and regulated to the plumbing industry, including the installation, alteration, repair or removal of plumbing and drainage systems.

Plumbing Fixtures. Plumbing fixtures are approved receptacles, devices, or appliances that are supplied with water and discharge wastes into a drainage system by a direct or indirect connection. Industrial and commercial processing vats, tanks, and similar equipment are not defined as plumbing fixtures, but may be discharged into approved fixtures or connected to the drainage system as directed by code.

Plumbing Official. The plumbing official is the official designated to lawfully enforce the plumbing code as adopted by a lawful government agency.

Potable Water. Potable water is water that meets the requirements of the health department having jurisdiction as fit for drinking water and other domestic uses.

Private or Private Use. Private use means that plumbing fixtures are intended solely for use of an individual or a single family in a residence, apartment, hotel, or similar establishments.

Private Sewage Disposal System. Private sewage disposal systems are septic tanks and drain fields located on private property and not considered as a public sewer or disposal system.

Private Sewer. A private sewer is a privately owned sewer.

Public or Public Use. Public use applies to any building, structure, establishment, recreation or works where plumbing fixtures are installed for unrestricted use of the public regardless of conditions of free or paid use.

Receptor. A receptor is a plumbing fixture or device approved for use to receive indirect wastes.

Relief Vent. A relief vent is a vent installed as an intermediate vent located at specified intervals to provide adequate circulation of air.

Rim. A rim is the unobstructed open edge of a plumbing fixture.

Riser. A riser is a pipe extending vertically through one or more stories, conveying water or gas to fixtures or appliances.

Roughing-In. Roughing-in is the placing or installing of all piping pertaining to the plumbing system prior to covering or installation of fixtures.

Seepage Pit. A seepage pit is a lined excavation used in conjunction with a septic tank designed to permit effluent to seep through the bottom and sides of the liner.

Septic Tank. A septic tank is a watertight tank designed to retain and digest solids and allow the liquids to discharge into the soil by seepage pits or open-pipe seepage lines.

Sewage. Sewage is any liquid waste containing matter foreign to water which must be disposed of through a drainage system.

Sewerage. Sewerage is the process and system that is required to carry off the wastes of entire community.

Single-family Dwelling. A single-family dwelling is a home occupied by one family on a single lot.

Slope. Same as grade.

Special Wastes. Special wastes are wastes that require special means of handling, not by ordinary domestic drainage requirements.

Stack. A stack is a vertical main waste or vent pipe extending through one or more stories.

Sump. A sump is an approved tank or pit that receives the discharge of fixtures below the level of a gravity drainage system that is emptied by pumps or mechanical means.

Supports. Supports means the material and method used to properly anchor, secure, or support the plumbing system or fixtures.

Trap. A trap is a fitting or device so constructed to retain a water seal to prevent a back passage of air or sewer gas without materially affecting the flow of waste or sewage through it.

Trap Seal. The trap seal is the amount of liquid retained in a trap between the top of the dip of the trap and its crown weir.

Vacuum Breaker. *See* backflow preventer.

Vent Pipe. A pipe or system of pipes to provide circulation of air within a drainage system to prevent back pressure and trap siphonage.

Vertical Pipe. A vertical pipe is any position or angle of installation not considered or defined as horizontal.

Waste. Waste is liquids entering the drainage system not containing fecal matter.

Water Distributing Pipe. Water distributing pipe is the water system supply to plumbing fixtures and other outlets and is potable water.

Water Main. Water main is the water pipe supply system for public or community use.

Water Service Pipe. Water service pipe is the pipe from the main, meter, or other source of supply to the building served.

Water Supply System. Water supply system is the complete water piping system from the source of community supply to all building premises outlets.

Wet Vent. A wet vent is a vent that may also act as a waste.

Yoke Vent. A yoke vent is a pipe connecting upward from a soil or waste stack to a vent stack for the purposes of preventing pressure changes in the stack.

Length of effective thread on pipe.
(Distance that is required to make a tight joint)

Nom. Pipe Size Inches	Dimension A Inches	Nom. Pipe Size Inches	Dimension A Inches
⅛	.2639	4	1.3000
¼	.4018	5	1 4063
⅜	.4078	6	1 5125
½	.5337	8	1 7125
¾	.5457	10	1 9250
1	.6828	12	2 1250
1¼	.7068	14 O.D.	2 2500
1½	.7235	16 O.D.	2 4500
2	7565	18 O.D.	2 6500
2½	1 1375	20 O.D.	2 8500
3	1 2000	24 O.D.	3 2500
3½	1 2500		

Dimensions given do not allow for variation in tapping or threading.

Distance "A" is the required length of entry to make a tight joint.

DISHWASHER INSTALLATION

1. Roughing-in. Install an air-gap fitting in a wall above a sink or drainboard. Install air-gap fitting to enter sink tail-piece connection or rough in separately. Where garbage disposal units are installed, enter disposal in connection provided in head. When disposals are installed, make sure that they are approved for a dishwasher connection. Where sinks and disposals are installed in one unit, always enter the disposal—not the sink tailpiece.

2. Water Connections. Provide hot water supply as directed by manufacturers' specifications.

3. Supports. Check with manufacturer's specifications.

4. Finish. Secure to wall and floor as required. Install hot water connections to washer inlet, provide shutoff valve. Install drain tube from pump to washer outlet. Connect air-gap drain to disposer head. When roughing in separately, install 1½ inches trap to the fixture tee. Where no disposal or separate rough-in is provided, install drain from air-gap fitting to Y directional tail-piece. When entering disposals, install short piece of approved flexible hose or rubber connection provided by manufacturer (not to exceed 6 inches).

5. Level and check for leaks.

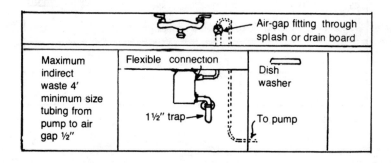

COMMERCIAL DISHWASHER

Most commercial dishwashers require a grease trap or interceptor. Locate grease trap within 4 feet of dishwasher. Protect all water inlets with vacuum breakers where required.

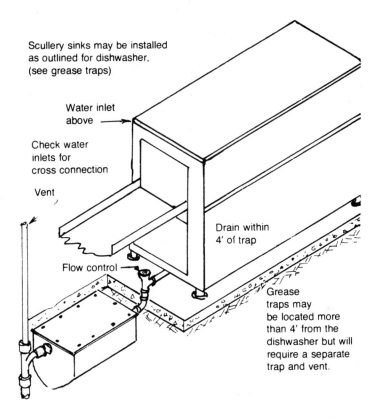

Scullery sinks may be installed as outlined for dishwasher, (see grease traps)

Water inlet above

Check water inlets for cross connection

Vent

Drain within 4' of trap

Flow control

Grease traps may be located more than 4' from the dishwasher but will require a separate trap and vent.

Scullery sinks may be installed as outlined for dishwashers (see grease traps).

All grease traps should be sized to properly take care of the grease requirements of the contemplated installation. Trap and vent all fixtures that exceed 4 feet from grease trap.

Where more than one fixture is served by one grease trap, size accordingly. On high temperature wastes locate trap for proper separation.

DOMESTIC DISHWASHER INSTALLATION

All such fixtures are required to waste to an indirect connection. Gravity wastes are not allowed. Install air-gap sink or drainboard.

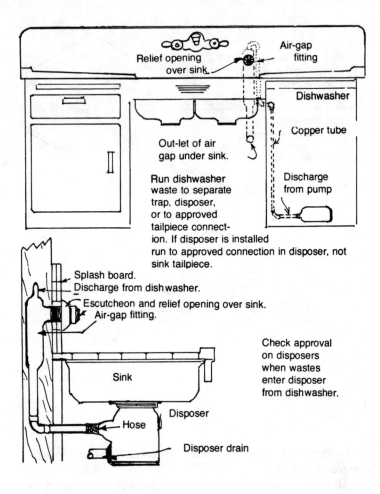

Relief opening over sink.

Air-gap fitting

Dishwasher

Copper tube

Out-let of air gap under sink.

Discharge from pump

Run dishwasher waste to separate trap, disposer, or to approved tailpiece connection. If disposer is installed run to approved connection in disposer, not sink tailpiece.

Splash board.
Discharge from dishwasher.
Escutcheon and relief opening over sink.
Air-gap fitting.

Check approval on disposers when wastes enter disposer from dishwasher.

Sink

Disposer

Hose

Disposer drain

DOUBLE SANITARY TEES FOR WASTE FITTINGS

Double sanitary tees are not recommended for wastes entering drains at same level. Use fittings so designed and constructed to prevent jumping across and entering the opposite waste. When not designed for this condition, increase the barrel of the fitting two pipe sizes.

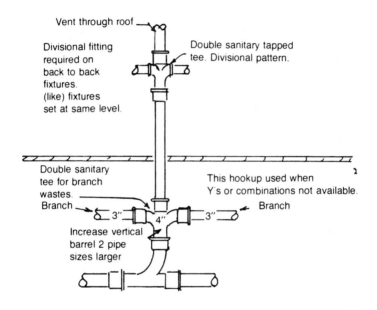

Vent through roof

Divisional fitting required on back to back fixtures. (like) fixtures set at same level.

Double sanitary tapped tee. Divisional pattern.

Double sanitary tee for branch wastes.

This hookup used when Y's or combinations not available.

Branch

Branch

3″   4″   3″

Increase vertical barrel 2 pipe sizes larger

A double sanitary tee is never used as a horizontal waste fitting. It may be used as a vertical fitting under conditions outlined by code. Consult your local inspection department if in doubt.

DRAWING A PIPING DIAGRAM

The most common method of drawing plumbing piping systems is the isometric shown below. Isometric lined paper can be obtained that will give a true 30 or 40 degree angle. Notice that all fittings can be shown with the isometric method, making a material list very easy to figure. Only 30-degree—60-degree triangles make a true isometric drawing.

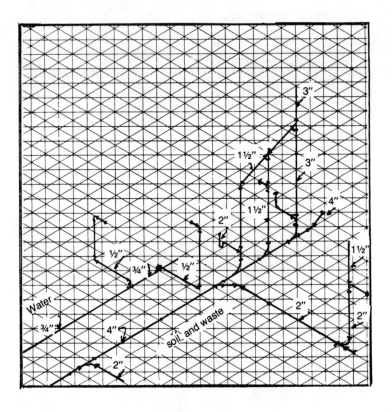

OBLIQUE DRAWING DIAGRAMS

A method of drawing a piping diagram is shown below. The main is projected along a 45-degree angle and the branches are drawn as shown.

Fittings, measurements, and sizes are marked on sketch showing how the installation is cut out and made up.

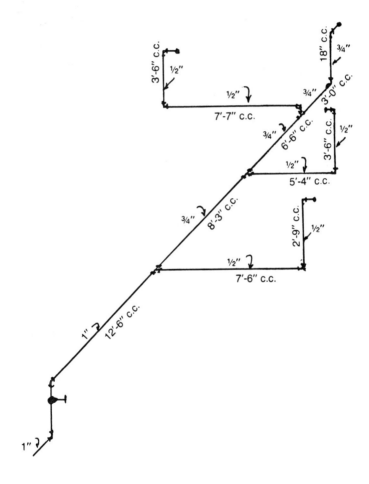

When illustrating a piping diagram all fittings, sizes and measurement of pipe should be clearly shown.

DRINKING FOUNTAIN WALL HUNG
AND STALL; CUSPIDOR OR DENTAL UNITS

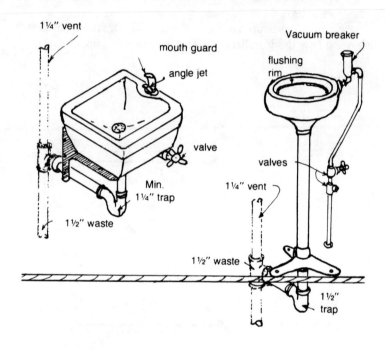

Drinking fountains and cuspidors or dental units may have 1½ inches vents and 1½ inches wastes. All drinking fountains must have an approved mouth guard and angle stream jet. All cuspidors require an approved vacuum breaker installed according to code requirements and health regulations. Drinking fountains are prohibited in toilet rooms.

DRAINAGE BELOW STREET LEVEL

Note: Size all building drains to allow for increased load from ejector. Figure two units for each GPM.

All back-water valves and gate valves must be installed where readily accessible. Dual pumps recommended on ejectors. Some localities require a trap where wastes enter sewer.

Separate venting for below-sewer installations required in some localities (check requirements). On air ejectors run exhaust separate. Subsoil drain sumps. Clear water drainage may enter receptor as shown. If hooked to sewer, provide check and gate valve.

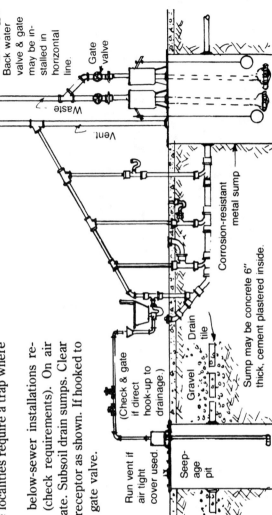

Enter waste ABOVE drain

Back water valve & gate may be installed in horizontal line.

Gate valve

Waste

Vent

Corrosion-resistant metal sump

(Check & gate if direct hook-up to drainage.)

Gravel Drain tile

Sump may be concrete 6" thick, cement plastered inside.

Run vent if air tight cover used.

Seepage pit

CHIMNEYS AND FLUES

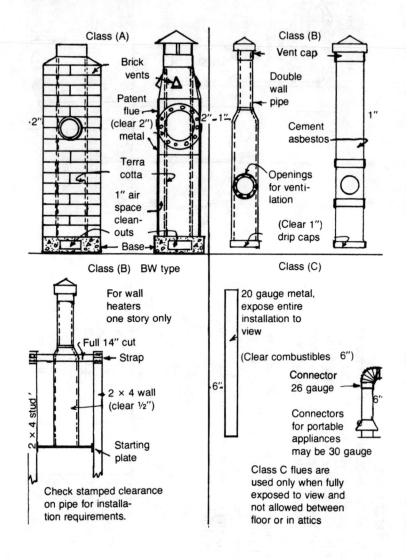

Class (A)

Brick vents

Patent flue (clear 2") metal

·2"

Terra cotta

1" air space clean-outs

Base

2"-1"

Class (B)

Vent cap

Double wall pipe

Cement asbestos

Openings for venti-lation

(Clear 1") drip caps

1"

Class (B)   BW type

For wall heaters one story only

Full 14" cut

Strap

2 × 4 wall (clear ½")

2 × 4 stud'

Starting plate

Check stamped clearance on pipe for installa-tion requirements.

Class (C)

20 gauge metal, expose entire installation to view

(Clear combustibles   6")

·6"

Connector 26 gauge

6"

Connectors for portable appliances may be 30 gauge

Class C flues are used only when fully exposed to view and not allowed between floor or in attics

62

VENT MATERIALS AND INSTALLATION OF FLUES

Class A. This flue is for high-heat appliances and may be of brick, masonry, lined masonry metal, terra-cotta patent type or designed to resist heat of over 550 degrees. All solid-fuel or oil-burning equipment requires class A flues.

Class B. This flue serves appliances that do not exceed 550 degrees of stack temperature. Materials may be cement asbestos, double-walled metal, or unglazed tile not less than ½ an inch thick.

Class BW. This flue is B class but has special clearances and is recommended for one-story installations only. Check for conditions.

Class C. This flue is a single wall metal flue of not less than 20-gauge galvanized iron or 24-gauge copper. Must be exposed installation.

CLEARANCES OF FLUES

Class A. Clear all combustibles 2 inches minimum. Where high heat appliances such as boilers and similar installations are encountered, check for additional requirements.

Class B. Clear all combustibles a minimum of 1 inch. Check all flues for stamped clearance.

Class BW. Clear ½ inches minimum front and back of wall. Cut the top plate a full 14½ inches to allow for a full ventilation. This installation is limited to one-story installations.

BW type flues are designed especially for wall heaters. Check stamped clearance label for wall heaters and flue installations.

SUPPORT OF FLUES

Class A. Support on foundation of masonry if of brick or masonry. (Reinforced concrete for brick chimneys or equal weight.) Patent flues to start on floor of structure, not attached to walls.

Class B. Support on brackets hung or installed with the building structure. Class BW. Supported to the appliance starting plate.

Class C. Support to the appliance or to starting plate.

INLETS TO FLUES

All flues enter Tee inlets unless especially designed for a vertical installation. Check for requirements if vertical flues are shown.

When entering flues, stagger inlets so that one inlet is above the other. If entering a manifold, when allowed, install Y connections. (See water heater manifold illustration.)

TERMINATION OF FLUES

Class A. Terminate not less than 2 feet above and 10 feet away from any portion of structure wall or roof of 45 degrees or more upward from the horizontal.

Gas Vents. Terminate at least 1 foot above and 4 feet away from any window, door, ventilator, air shaft, or air intake. Clear vertical walls or portions of the structure within 4 feet of vent. (See chimneys and flues.) Clear portions of structure 45 degrees or more upward from the horizontal.

ATTIC FURNACE INSTALLATION

1. Provide Permanent, Continuous Access to Furnace Controls. Access door must be large enough to accommodate the removal of the largest single piece or part of the furnace. A catwalk from the access door should be at least 30 inches wide run continuous to furnace controls.

2. Base. Set furnace on solid fireproof base according to AGA clearances as noted on plate. When setting on combustible floors, fireproof as required by building official. Base should extend beyond furnace at least 12 inches in all directions. Locate furnace over hallway if possible. Do not install where ceiling joists are not designed to carry the additional load.

3. Gas Connections. Install plug type main gas shutoff within 3 feet of furnace controls. If furnace is rated at 100,000 Btu or over, install rigid pipe hookup. Appliance connectors may be used if under 100,000 Btu rating. Gas pipe size must be checked for load demand.

4. Name plate. Check name plate for rating and type of gas. Do not attempt to install an appliance marked for LPG where natural gas is supplied for use. Follow clearances as stated on name plate for protection to combustibles.

5. Flue. Install T inlet and flue support as required. Check clearances to wood. Allow at least 1 inch clearance on all B type flues. Run flue 12 inches through roof and not closer than 4 feet to any door, window, vertical wall, air intake, or ventilating shaft. Install approved roof jack and vent cap. Do not project flue into vent cap outlet; allow for full ventilation.

6. Vent. Class C vent connectors must clear combustibles by at least 6 inches. Metal screw all joints. No black stove pipe allowed.

7. Plenum and Hot Air Ducts. Fire-proof all combustible materials within 9 inches of plenum and 3 inches of hot air ducts. Cover pipe where required.

8. Cold Air Returns to be in Ducts of Noncombustible Materials. No opening in cold air or re-circulated air within 10 feet of draft diverter or flue exhaust.

9. Combustion Air. All furnaces must be supplied with sufficient air to properly support combustion. Make sure there is ample combustion air.

10. Light. Install switch-operated light at or near furnace controls.

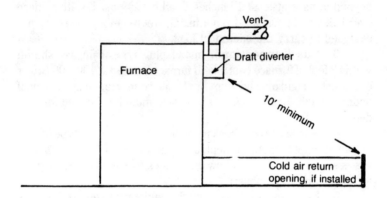

Where attic furnaces are located, be sure that there is ample support. Install to rigid fire protection clearances on attic furnaces. Locate all valves and controls where readily accessible. Check attic furnaces at frequent intervals.

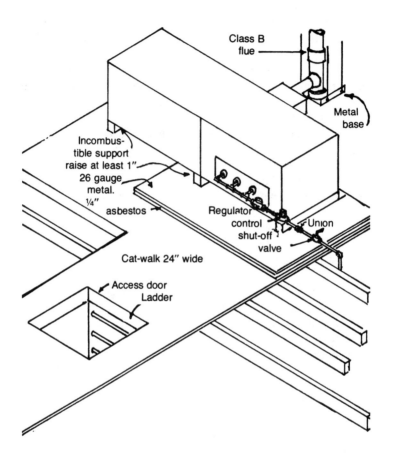

Class B flue

Metal base

Incombustible support raise at least 1". 26 gauge metal.

¼" asbestos

Regulator control shut-off valve

Union

Cat-walk 24" wide

Access door
Ladder

Provide light and switch. Access door must be within 20 feet of furnace controls, and large enough to remove largest part of furnace.

Provide sufficient air to properly support combustion. Install main gas shut-off in addition to control valve supplied with appliance.

CENTRAL FURNACE INSTALLATION

1. **Room Size.** Must be large enough to properly service or repair any part of furnace. Provide 18 inches clearance to furnace controls. Install in accordance with clearances as approved by A.G.A.

2. **Combustion Air.** Allow at least 1 square inch ventilation for each 1000 Btu rating. Air to be supplied in ducts to outer air in one or more in top and one or more in bottom of furnace room. Install screen over openings; ¼ inches mesh.

3. **Circulating Air.** Return air in tight ducts in furnace cold air plenum. Filters must be installed in accessible location for ready removal.

4. **Ducts.** Must be sized to adequately heat the building in which they are installed. Insulate all ducts within 3 inches of combustible materials. Cold air returns must equal in cross sectional area warm air supply. Cold air ducts may be formed by materials allowed by code other than metal if they are made of one-hour fire-resistant materials. All ducts shall be impervious, smooth, and leak proof.

5. **Re-circulated Air.** Return air openings must be located where there is no possibility of return into the furnace of fumes, objectionable odors, or harmful air. Keep 10 feet from diverter box.

6. **Vent Connector.** Allow 6 inches clearance to combustible materials. Must not be larger than flue.

7. **Flue.** Install in vertical position at least 1 foot through roof and 4 feet from any vertical wall, window, door, or air intake.

8. **Fireproofing.** Install to clearances specified by plate mounted on furnace.

9. **Plenum.** Unless constructed and approved for specified clearance, protect all combustible materials within 6 inches. For reverse flow use A.G.A. or approved floor connection.

10. **Gas Connection.** Rigid hookups required on all appliances rated over 100,000 Btu. Provide approved main shutoff in accessible position near furnace controls. Install vent relief tube from gas regulator to pilot light if regulator designed and tapped for such. Terminate relief at and ½ inches below the pilot light.

Do not install in bathroom or bedroom or any closet directly off such rooms. Use rigid hookup if appliance rated 100,000 Btu or over. Protect combustibles where plenum or heat pipes pass through vent connection.

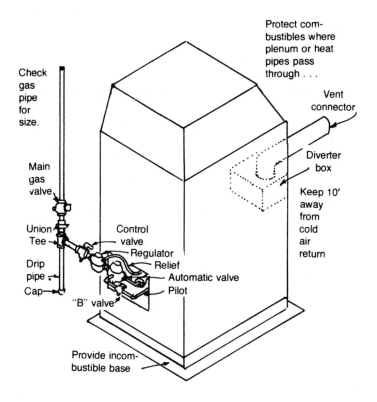

Where down-flow installations plenums pass through combustible floors, check for A.G.A. requirements. Double combustion air for L.P.G. Provide 1 square inch per BTU combustion air to outer air at floor and ceiling.

FLOOR FURNACE INSTALLATION

1. **Location.** There is a 20-foot maximum from access to control valves of furnace, burners or operating parts.

2. **Clearance.** There is a 24-inch minimum cross-sectional area of crawl trench or under building.

3. **Wet Areas.** Not allowed unless fully protected against seepage.

4. **Pit.** Must be 1 foot larger than furnace in any dimension or direction and 18 inches larger on control side. Dig 6 inches deeper than lowest point of burner.

5. **Flue.** Provide flue size required by furnace. Enter flue through T inlet, supported on bracket hung to building structure. (Do not start flue on ground or pier.) Provide clearances as required for type of flue used. Terminate as required. Vent. Use proper weights as specified and maintain clearances as required. (Black stove pipe not allowed.) Metal screw all connections. When installing a diversion box, make sure that it is right side up. Secure vent to diversion box.

6. **Pitch.** Pitch up 1 inch per foot, and never less than ½-inch under any condition.

7. **Horizontal Length.** Not to exceed three-quarters of the total vertical length.

8. **Fireproofing.** Fireproof all flues, vents, down-draft diverters closer to combustibles than approved with an approved air space of ½ inch asbestos covered with 26-gauge metal.

9. **Hangers.** Hang to building structure members. Do not build up supports from the ground.

10. **Gas Connections.** Check label for type of gas and size of supply. Where connectors are used, install only the approved type. Provide a drip cap at low point of gas supply. (Only ground joint unions are allowed when unions are installed on gas lines.) Install proper approved gas cocks. Make sure that the appliance can be disconnected when gas cock is turned off.

11. **Grills or registers.** Locate 6 inches away from doors or walls. Doors must not open over grill.

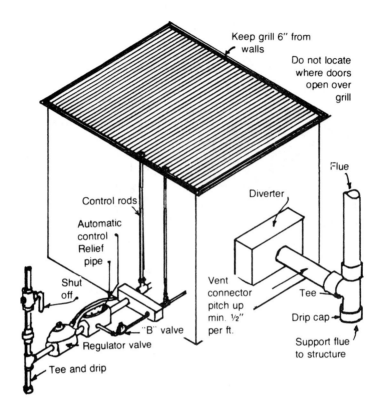

Keep grill 6" from walls

Do not locate where doors open over grill

Flue

Diverter

Control rods

Automatic control

Relief pipe

Shut off

Vent connector pitch up min. ½" per ft.

Tee

Drip cap

"B" valve

Regulator valve

Support flue to structure

Tee and drip

Cut a hole or install a header ¼ inch larger than actual dimension of floor furnace. Dig a pit 6 inches below the lowest part of furnace. Dig a pit 12 inches larger than the furnace. Provide 24 inches access within 20 feet of controls. Locate access to furnace must be 24 inches in any cross-sectional area.

Do not install a floor furnace in wet locations or where flooding can occur.

FLOOR FURNACE INSTALLATION (seepage areas)

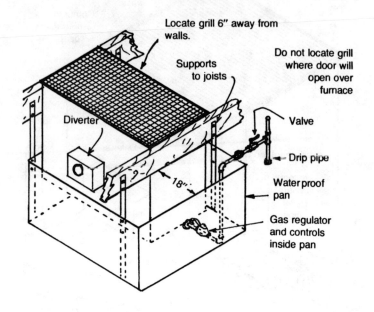

Locate grill 6″ away from walls.

Supports to joists

Do not locate grill where door will open over furnace

Diverter

Valve

Drip pipe

Water proof pan

Gas regulator and controls inside pan

Install corrosion-proof pan where danger of flooding exists. Pipe to enter above top of pan. Locate access hole to furnace within 20 feet of furnace. 24 inches clearance required to furnace in all directions. Top of pan at least 2 inches above outside natural grade. Extend pan 18 inches larger than furnace in front. Support pan to joists above to prevent water pressure lifting pan against furnace.

WALL HEATER INSTALLATION

1. **Vent.** Install by clearances required for types of vent. In 4-inch walls, use "BW" installation. Cut out entire plate full 14 inches. Install centering plate to provide for ½-inch clearance to walls. When using class B vents, maintain full 1 inch clearance to combustible materials.

2. **Strapping.** Where plates are cut through, install metal straps one-eighth inch thick and 22 inches long, nailed with 16-penny spikes at each end.

3. **Flue.** Extend up through roof into approved roof cap not less than 12 inches above roof. Do not terminate vent within 4 feet above vertical wall, door, window, vent shaft, or air intake.

4. **Flue Cap.** Install approved type flue cap. When installing be sure that flue does not project into cap ventilating area.

5. **Gas Connections.** Install approved type main gas shutoff. When installing tubing, use manufactured flared type only. Do not bend kinks into tubing. Kinked bends reduce the flow of gas into the appliance from the supply.

6. **Location of Controls.** Run rigid pipe through floor into burner area. Gas shutoff must be installed in accessible area near controls.

7. **Gas Regulator.** Install relief tube to pilot light when required.

8. **Appliance Approval.** Check for name plate stating type of gas, burner rating and required clearances.

9. **Piping Under Floors.** Run rigid pipe under floors through walls or in concealed places. Tubing connectors are allowed in exposed places and are limited to 3 feet in length.

10. **Automatic Controls.** Install so that when controls are shut off pilot light will remain open.

11. **Location of Thermostat.** Install room thermostat according to manufacturer's recommendations. A poorly located thermostat will not operate the appliance control properly.

WALL FURNACES (in 2 × 4 walls)

Terminate vent 4 feet away and 1 foot above any door, window, air intake or opening into building. Extend above any wall or section of building built at angle of 45 degrees upward from the horizontal.

This installation approved for one story with 2-×-4 wall.

Cut between studs must be 14 inches or more to allow for ventilation.

On two-story work, consult Inspection Department.

Rigid gas connection through floor. Install approved gas heater in accessible position.

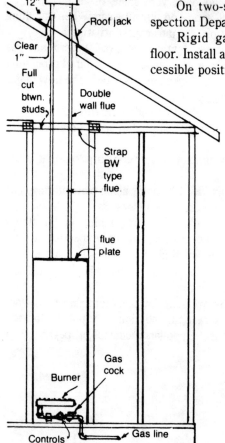

Clear 12″
Vent cap
Roof jack
Clear 1″
Full cut btwn. studs
Double wall flue
Strap BW type flue.
flue plate
Gas cock
Burner
Controls
Gas line

GARAGE WASH RACK INTERCEPTORS

Minimum size of primary interceptor 18 inches wide 36 inches long and 24 inches deep.

Minimum size of secondary interceptor 18 inches wide 18 inches long and 24 inches deep.

Elevate secondary interceptor above grade to prevent it acting as a floor drain. No cleanouts permitted inside secondary interceptor. Maximum separation between interceptors 25 feet unless separately vented. Provide deep seal traps on all interceptors.

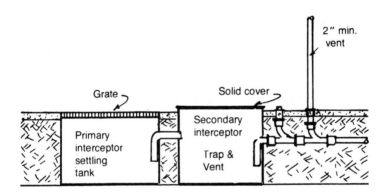

Minimum construction 4 inches concrete, cement plaster inside ½ inch thick. Minimum drain size 3 inches for interceptors. Minimum vent size 2 inches. Check with local inspection department for specifications for tanks and sand traps.

GAS DRYER INSTALLATION (DOMESTIC)

1. Piping. Rigid piping required through floor, walls or concealed places. Tubing connectors allowed if not over 3 feet long and accessible.

2. Shut-offs. Install approved main shut-off in accessible position.

3. Approvals. Check for approval plate mounted on appliance stating type of gas, Btu rating, and required clearances.

4. Location. Locate on outer wall if possible. No dryers in bath or bedrooms. If installed in garage, protect against mechanical injury.

5. Vent. Vent all gas appliances to outside. When located on outside wall, use approved terminal furnished for appliance. Do not exhaust under house area. When located on interior wall vent as allowed by ordinance.

6. Clearances. Vents of class C construction must have 6 inches clearance to combustibles. Maintain 1-inch clearance on class B vent.

7. Combustion air. If dryer is installed in rooms of 25 square feet or less, provide fixed ventilation top and bottom not less than 36 square inches for each ventilator. Total 72 square inches.

8. Controls. Provide relief tube to pilot light when required.

9. Setting. Set appliance on firm level floor, and adjust legs or adjusting bolts to ensure against rocking or vibration.

10. Do not attempt to vent a gas dryer into any other appliance vent. Take care to clean lint after usage. Do not locate the gas dryer in unventilated spaces. Do not exceed factory design specifications. Consult with your inspection department when in doubt.

11. Gas dryer located in garage should be installed where a vent can be provided. Consult with your local inspection agency as to requirements regulating gas appliances in private garages.

GAS DRYERS

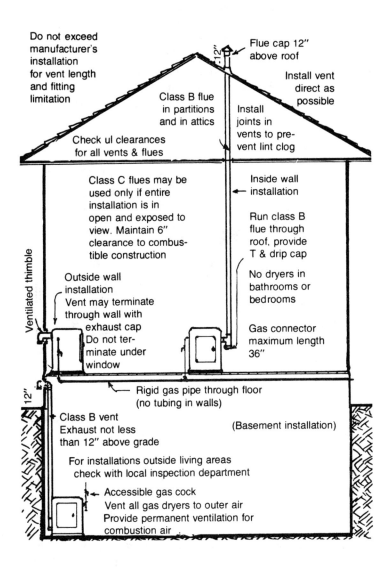

Do not exceed manufacturer's installation for vent length and fitting limitation

Flue cap 12" above roof

Install vent direct as possible

Class B flue in partitions and in attics

Install joints in vents to prevent lint clog

Check ul clearances for all vents & flues

Class C flues may be used only if entire installation is in open and exposed to view. Maintain 6" clearance to combustible construction

Inside wall installation

Run class B flue through roof, provide T & drip cap

Ventilated thimble

Outside wall installation
Vent may terminate through wall with exhaust cap
Do not terminate under window

No dryers in bathrooms or bedrooms

Gas connector maximum length 36"

12"

Rigid gas pipe through floor (no tubing in walls)

Class B vent
Exhaust not less than 12" above grade

(Basement installation)

For installations outside living areas check with local inspection department

Accessible gas cock
Vent all gas dryers to outer air
Provide permanent ventilation for combustion air

GAS AND ELECTRIC METER BOX

Gas and electric meters may be installed in one box if separated by a gas-tight partition. Minimum size gas recess is 30 inches wide, 18 inches deep and 26 inches high.

Minimum size recess for electric meter is 12 inches wide, 12 inches deep and 26 inches high. Consult utility company when installing gas and electric meters in one recess. (Allow at least 30 inches under meter box for gas company header or manifold.)

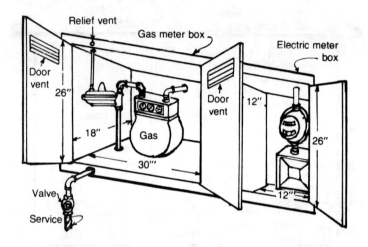

Keep bottom of box up 30 inches from finished grade. Rough out house stub 4 inches from wall. Keep house supply up 20 inches.

Meter boxes and valves must be accessible at all times. When enclosed gas meters are installed, a ventilated door must be provided.

Relief vents should be provided to the outer air and 5 feet from any window or ventilator.

When installing multiple meters provide identification tags for all meters.

Gas meters are prohibited inside living quarters, closets, restrooms, boiler rooms, under stairways, or in unventilated locations.

GAS METER INSTALLATION

Gas meters are prohibited under interior stairways, boiler, engine, heater, electric meter rooms or under show windows and spaces that are not readily accessible for reading, inspection or repair. When installed in a box, see local utilities for regulations. Minimum box should not be less than 22 × 24 inches. Rough apartment-house gas lines 12 inches to right and 20 inches from ground.

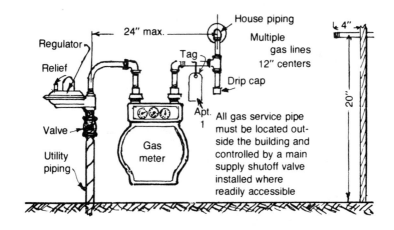

All meters should be located at one place unless impractical to do otherwise. When more than one meter is installed at the same location, all meters must be identified with a permanent tag attached to the meter. When in doubt, consult the local utility company for regulations and information on location of meter.

Normally gas roughs in to the right of the service with 24 inches. (When in the open and if allowed, gas meters may be installed under exterior or outside stairways if accessible (check this.) Terminate reliefs 5 feet from ventilation openings.

OUTDOOR GAS LIGHTS

If not approved for reduced clearances on light, clear all combustible walls, ceilings, or parts of all structures at least 18 inches in any direction.

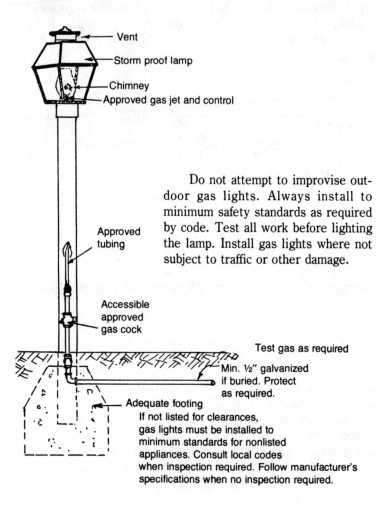

Vent

Storm proof lamp

Chimney

Approved gas jet and control

Do not attempt to improvise outdoor gas lights. Always install to minimum safety standards as required by code. Test all work before lighting the lamp. Install gas lights where not subject to traffic or other damage.

Approved tubing

Accessible approved gas cock

Test gas as required

Min. ½″ galvanized if buried. Protect as required.

Adequate footing

If not listed for clearances, gas lights must be installed to minimum standards for nonlisted appliances. Consult local codes when inspection required. Follow manufacturer's specifications when no inspection required.

GAS LOG LIGHTERS

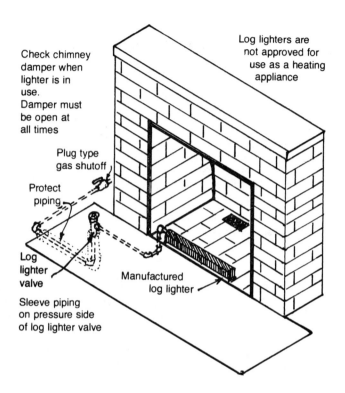

Check chimney damper when lighter is in use. Damper must be open at all times

Log lighters are not approved for use as a heating appliance

Plug type gas shutoff

Protect piping

Log lighter valve

Manufactured log lighter

Sleeve piping on pressure side of log lighter valve

All gas lighters in a fireplace must be an approved device that is manufactured and designed for this use. The log lighter valve must be an approved valve, located in the same room, and not more than 4 feet from the device. An approved shutoff valve should be installed to control the log lighter valve and be where it is readily accessible. Piping through masonry must be galvanized or noncorrosive and be covered by at least 2 inches of masonry.

GAS PIPE INSTALLATION

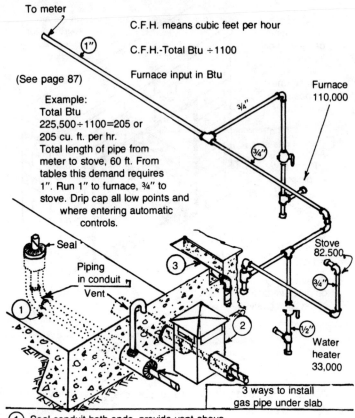

To meter

(See page 87)

C.F.H. means cubic feet per hour

C.F.H.-Total Btu ÷1100

Furnace input in Btu

Example:
Total Btu
225,500÷1100=205 or
205 cu. ft. per hr.
Total length of pipe from
meter to stove, 60 ft. From
tables this demand requires
1″. Run 1″ to furnace, ¾″ to
stove. Drip cap all low points and
where entering automatic
controls.

Furnace
110,000

Stove
82.500

Water
heater
33,000

Seal

Piping
in conduit
Vent

3 ways to install
gas pipe under slab

(1) Seal conduit both ends, provide vent above
ground level.
(2) Seal conduit through floor, leave end open
into box. Extend ventilated box 6″ above
ground.
(3) Provide channel in floor, cover channel.

82

SIZING GAS LINES

Gas tables are carefully engineered to provide a safe supply of gas for a predetermined length of pipe. This supply is according to size and length and will deliver so many cubic feet of gas per hour, (CFH) allowing for pressure drop and demand factors. Note: Stay with same length for all figures including branches. Use only greatest length.

1. Determine greatest length of pipe from meter to most remote outlet.

2. Add input Btu rating of all appliances combined. (Convert C.F.H.)

3. Knowing length in feet, find in pipe size column size of pipe to deliver required CFH.

4. After finding pipe size required at meter, which combines all demands, proceed to find pipe size for outlets.

5. Refer to preceding page.

Length of line = 60 feet
Total demand in CFH = 205
Line size --------- 1 inch

*Stove @ 82,500 Btu *Min. ¾" to any stove	$\dfrac{82,500}{1100}$	= 75 CFH
Water heater @ 33,000 Btu	$\dfrac{33,000}{1100}$	= 30 CFH
Furnace @ 110,000 Btu	$\dfrac{110,000}{1100}$	= 100 CFH 205 CFH total

6. Staying in the 60-foot length and working from the stove to the meter, we find ½ inch too small so ¾ inch is selected. Adding the stove at 75 and the water heater at 30 CFH, we find that ¾ inch is still large enough. Adding in the furnace at 100 we can find a total of 205 CFH, which will require a 1 inch pipe at this point. Note: If gas regulators are installed, start sizing low-pressure lines from the regulator.

GAS PIPE INSTALLATION

1. Demand. The demand in cubic feet per hour is determined from the input rating of the appliance. Each appliance or device must have an identification showing Btu input, clearances, and other pertinent information.

2. To Determine CFH. Total all Btu input ratings and divide by 1100.

3. Length. Measure to the most remote outlet to find applicable length from tables.

4. Knowing Length and CFH. From tables find the pipe size it will take to deliver the demand required. This will give size at meter.

5. Staying in Same Length Column. Work backward toward the meter and add appliance demands. When CFH demand exceeds, pipe size increases accordingly.

Appliance Ratings

Appliance	Demand
Domestic gas range	75 CFH
Domestic recessed top burner section	50
Domestic recessed oven section	25
Storage water heater up to 30 gallons	30
Storage water heater 40 to 50 gallons	45
Domestic clothes dryer	20
Fireplace log lighter residential	25
Fireplace log lighter commercial	50
Barbecue residential	50
Gas refrigerator	3
Bunsen burner	3
Gas engines (per horsepower)	10
Steam boilers (per horsepower)	50
House trailers (mobile homes) first outlet on any branch or main	125
second outlet	100
third outlet	75
Subsequent loadings after third outlet	50

RECOMMENDED PIPE SIZES (STANDARD PIPE SIZE)

Length in feet	Flow of gas in pipes with a pressure drop of ½″ water column. Specific gravity .64 Pipe capacity in CFH. below pipe size column							
	½″	¾″	1″	1¼″	1½″	2″	2½″	3″
10	95	275	580	1050	1675	3550	6350	10200
20	65	187	398	720	1150	2420	4350	6950
30	52	152	320	580	920	1950	3500	5590
40	44	130	274	495	790	1670	3000	4780
50	39	115	242	438	700	1480	2650	4200
60	36	105	220	396	630	1390	2400	3800
70	33	96	202	365	580	1280	2210	3500
80	31	89	188	340	540	1200	2050	3280
90	29	84	177	319	510	1075	1930	3030
100	27	79	167	300	480	1020	1830	2900
125	24	70	148	265	423	900	1630	2570
150	22	63	134	240	382	810	1480	2340
175	20	58	123	220	352	750	1360	2150
200	18	54	115	205	329	695	1270	2000
225	17	51	108	193	308	652	1190	1870
250	16	48	102	182	290	613	1150	1770
275	15	45	97	173	275	585	1070	1680
300	14	43	92	165	264	560	1020	1600
350	14	40	85	152	242	515	935	1480
400	13	37	79	142	225	480	870	1380
450	12	35	74	133	212	450	815	1290
500	11	33	70	125	200	422	770	1220
600	10	30	63	118	181	382	695	1100
700	9	28	58	105	167	351	640	1020
800	9	26	54	97	155	320	595	945
900	8	24	51	91	145	310	560	880
1000	8	23	48	86	138	290	529	830
Add—	1.6	2.2	2.7	3.5	4.3	5.2	6.4	8.0

For each standard elbow add to length of pipe as shown in above columns.

RECOMMENDED PIPE SIZES AND CAPACITIES IN CFH

Length in feet	Flow of gas in pipes with a pressure drop of ½" w.c. spec. gravity .64 — Pipe capacity in CFH below in pipe size col.					
	4"	5"	6"	8"	10"	12"
10	21750	39000	62000	132500·	237500	380000
20	15000	26700	42500	90000	163000	260000
30	12000	21500	34000	72000	130000	210000
40	10300	18400	29200	61500	112000	180000
50	9050	16300	28000	54300	99000	159000
60	8200	14700	23400	49100	90000	144000
70	7500	13600	21500	45300	82500	132000
80	7000	12600	20000	42000	77000	124000
90	6560	11800	18700	39500	72000	115000
100	6200	11200	17700	37200	68000	110000
125	5500	10000	15700	33100	60000	96000
150	4980	9000	14200	30000	54200	87000
175	4540	8300	13000	27600	50000	79500
200	4260	7700	12200	26500	46500	74000
225	4000	7150	11400	24200	43500	69300
250	3790	6850	10700	22800	41000	65500
275	3590	6500	10200	21600	39000	62000
300	3420	6200	9700	20700	37100	59100
350	3150	5700	8900	19000	34200	54500
400	2930	5300	8300	17700	32000	50400
450	2750	4970	7800	16500	30000	47500
500	2600	4700	7450	15600	28500	45000
600	2350	4250	6650	14200	25700	41000
700	2160	3900	6100	13000	23500	37500
800	2010	3620	5700	12200	22000	34900
900	1890	3400	5350	11500	20700	32800
1000	1790	3320	5030	10800	19500	31000
Add—	11.0	14.0	16.0	21.0	26.0	32.0

For each standard elbow add to length of pipe as shown in above columns.

GAS PIPE SIZING TABLES (ABOVE 8-INCH W.C.)

Clear all sizing with utility furnishing gas in each locality
(for initial pressure of 2 psi terminal pressure 1 psi)

Read cfh in column under Pipe Size Gas at 0.65 specific gravity

(for initial pressure of 2 psi terminal pressure 1 psi)

Length in feet	3/4"	1"	1¼"	1½"	2"	2½"	3"	4"	5"	6"
50	1050	2220	4000	6400	13700	24500	39500	85000	2 psi table	
100	720	1520	2700	4400	9400	16800	27000	58500		
250	445	940	1670	2720	5800	10200	16600	36000		
500	310	640	1140	1850	3950	7100	11300	24400		
1000	210	445	800	1280	2750	4900	7800	16800		

(for initial pressure of 3 psi terminal pressure 1.5 psi)

Length in feet	3/4"	1"	1¼"	1½"	2"	2½"	3"	4"	5"	6"
100'	1180	2200	4250	6750	12800	20700	38900		3 psi table see page 77B	
250	730	1350	2750	4100	8000	12700	23700			
500	500	940	1900	2890	5450	8700	16200			
750	405	750	1530	2300	4400	7000	13100	27200		
1000	345	650	1310	1975	3790	6000	11200	23300		
1500	275	515	1040	1570	3020	4800	9000	18700	31700	55700
2000	235	440	900	1340	2580	4100	7700	16000	27000	47000
2500	210	390	800	1190	2290	3650	6800	14200	24000	42000

(for initial pressure of 5 psi terminal pressure 1.5 psi)

Length in feet	3/4"	1"	1¼"	1½"	2"	2½"	3"	4"	5"	6"
100'	1930	3650	7300	11000	21000	33800	63000		5 psi table	
250	1180	2230	4500	6800	12800	20600	38700			
500	815	1520	3100	4650	8900	14300	26600			
750	660	1120	2500	3770	7200	11300	21400	44500		
1000	555	1030	2120	3180	6100	9750	18300	38200		
1500	450	840	1710	2570	4900	7800	14700	30500	52000	90000
2000	390	720	1470	2190	4200	6700	12500	26000	44000	77000
2500	340	640	1295	1950	3750	5950	11100	23100	39200	68000

SIZING GAS LINES COMBINING HIGH AND LOW PRESSURE

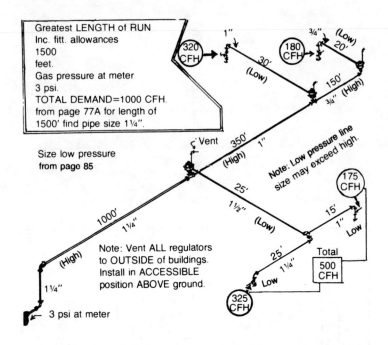

Greatest LENGTH of RUN
Inc. fitt. allowances
1500
feet.
Gas pressure at meter
3 psi.
TOTAL DEMAND=1000 CFH.
from page 77A for length of
1500' find pipe size 1¼".

Size low pressure
from page 85

Note: Vent ALL regulators
to OUTSIDE of buildings.
Install in ACCESSIBLE
position ABOVE ground.

Note: Low pressure line
size may exceed high.

    Installations combining high and low pressures must be con-
sidered as separate lines. Low pressure (8- to 11-inch water column) is figured from the gas regulator. High pressure is figured
from the meter to the last regulator. See previous page.

Total length of high pressure run -------------------------- 1500 feet
Total demand in cubic feet per hour------------------------------ 1000
High pressure pipe size required at meter---------------- 1¼ inches
Staying in 1500 foot range----------------------- size back to meter
Most remote demand requires for 180 CFH -------------- ¾ inches
Adding 180 and 320 = 500 CFH requires --------------------- 1 inch
Adding 500 + 325 + 175 = 1000 CFH requires-------- 1¼ inches
(Low pressure) demand of 180 @ 20 feet requires ------- ¾ inches
30 foot run @ 320 CFH requires --------------------------------- 1 inch
Combined total of 500 CFH for 50 feet requires------- 1½ inches
325 CFH requires 1¼ inches---------- 175 CFH requires -- 1 inch

88

GAS STOVE INSTALLATION

1. Location. Install to comply with minimum clearances as required by appliance approval.

2. Vent. Install vent where required to flue connection. Clear combustible materials 6 inches; protect as required by code.

3. Flue. See that the flue is installed properly and extends through the roof at least 12 inches. Terminate 4 feet from windows, doors, vertical walls, or air intakes. Flue must be as large as the appliance outlet and is unobstructed in its entire length. Install approved vent cap.

4. Nonvented Stoves. Where permitted, so-called non-vented stoves may be installed without the conventional vent, if an approved ventilating pipe is installed immediately above the stove and extends to the outer air. Minimum opening for this gravity ventilator is 6 inches by 8 inches, and the minimum pipe size from ventilator through roof is 6 inches.

5. Appliance Approval. Check for approval stamp stating type of gas and burner rating.

6. Gas Supply. Check stamp for Btu rating, and size gas supply accordingly. Do not supply appliance with a gas pipe smaller than the appliance inlet.

7. Gas Connection. Install approved gas cock in accessible position. When tubing connectors are allowed, check for approval and proper size.

8. Built-In Stoves. When built-in stoves are installed, follow the manufacturer's instructions and provide clearances as required by testing agency.

9. Used Stoves. Do not install stoves that have no identifying labels or laboratory approval seal. Do not supply gas to any stove unless the type of gas is clearly identified on the approval stamp.

10. Adjustments of Burners. Do not attempt to adjust burners unless you are thoroughly familiar with this type of work. Check with your local utility company or tradesman specializing in this field if the burner is out of adjustment.

11. Trash Burners. Where stoves are equipped with trash burners, class A flues must be provided. Do not install trash burners to a class B flue.

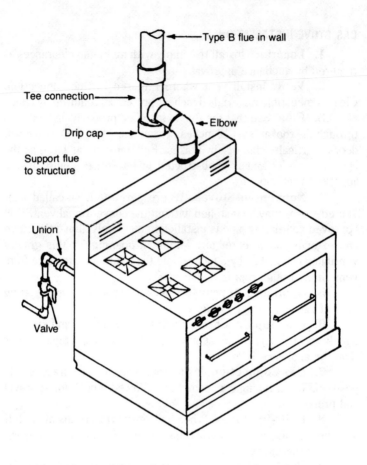

Minimum size of gas pipe three-quarter inch. Main shut-off valves must be accessible and approved type. Vent from stove must connect to an approved flue.

DEFINITIONS OF GAS TERMS

Air Mixer. That portion of an injection (Bunsen) type burner into which the primary air is introduced.

Air Shutter. An adjustable device for varying the size of the primary air inlet or inlets.

Appliance. A gas appliance is any device which utilizes gas fuel to produce light, heat or power.

Automatically Controlled Appliance. Appliances equipped with an automatic pilot and other automatic devices that: (a) Accomplish complete turnon and shutoff of the gas to the main burner or burners. (b) Graduate the gas supply to the burner or burners, but do not affect the complete shutoff of the gas.

Appliance Flue. The flue passages within the appliance.

Approved. "Approved" means acceptable to the authority having jurisdiction.

Automatic Gas Shutoff Valve. A device so constructed that the attainment of a temperature in the medium being heated in excess of some predetermined limit acts upon a chemical or metallic element in such a way as to cause the gas to the appliance to be shut off and remain shut off.

Automatic Ignition. Automatic ignition shall be interpreted as means which provide for ignition of the gas at a burner valve controlling the gas to that burner is turned on, and will effect re-ignition if the flames on the burner have been extinguished by means other than closing the gas burner valve.

Automatic Pilot. Consists of an automatic pilot device and pilot burner securely assembled in fixed functional relationship.

Atmospheric Injection Burner. A burner in which the air at atmospheric pressure is injected into the burner by a jet of gas.

Luminous or Yellow-Flame Burner. A burner in which secondary air only is depended on for the combustion of the gas.

Power Burner. A burner in which either gas or air or both are supplied at pressures exceeding, for gas, the line pressure, and for air, atmospheric pressure; this added pressure being applied at the burner.

Premixing Burner. A power burner in which all or nearly all of the air for combustion is mixed with gas as primary air.

Pressure Burner. A burner supplied with an air gas mixture under pressure (usually from 0.5 to 14 inches of water and occasionally higher).

Central Heating Gas Appliance. A vented gas-fired appliance comprising the following classes as defined herein: boiler, central furnace, floor furnace, or vented-recessed heater.

Chimney. A vertical masonry or reinforced concrete shaft containing one or more flues or vents.

Closed-Water Piping System. A system of water piping where a check valve or other device prevents the free return of water or steam to the water main.

Clothes Dryer. A device used to dry wet laundry by means of heat.

Combination Range. See dual oven combination of fuel gas.

Combustible Construction. A combustible wall or combustible surface constructed of wood, composition, or of wooded studding and lath and plaster.

Automatic Pilot. Device employed with gas-burning equipment that will either automatically shut off the gas supply to the burner(s) being served or automatically actuate, electrically or otherwise, a gas shutoff device when the pilot flame is extinguished. The pilot burner may or may not be constructed integrally with the device.

Automatic Pilot Device, Complete Shutoff Type. An automatic pilot device for automatically shutting off the gas supply to the main burner and pilot in event of pilot or gas failure.

Baffle. An object placed in the appliance to change the direction of or retard the flow of air, air gas mixture, or flue gases.

Boiler. A self-contained, gas-burning appliance for supplying hot water or low-pressure steam, primarily intended for domestic and commercial space-heating application.

Branch Line. Gas piping that conveys gas from a supply line to the appliance.

Btu. Abbreviation for British thermal unit that is the quantity of heat required to raise the temperature of 1 pound of water 1 degree Fahrenheit.

Bungalow (Utility). Type domestic gas range. A domestic range having a gas oven or top section or both, and a gas, electric, solid or liquid-fuel section designed for space heating or heating a solid top section, but not for oven heating.

Burner. A device for final conveyance of the gas, or a mixture of gas and air, to the combustion zone.

Injection (Bunsen) Burner. A burner employing the energy of a jet of gas to inject air for combustion into the burner and mix it with the gas.

Combustion. Combustion, as used herein, refers to the rapid oxidation of fuel gases accompanied by the production of heat or heat and light. Complete combustion of a fuel is possible only in the presence of an adequate supply of oxygen.

Combustion Chamber. The portion of an appliance in which combustion occurs.

Combustion Products. Constituents resulting from the combustion of a fuel with the oxygen of the air, including the inerts but excluding excess air.

Concealed Gas Piping. Gas piping, when in place in the finished building, would require removal of permanent construction to gain access to the piping.

Condensate (Condensation). The liquid that separates from a gas (including flue gas) due to a reduction of temperature.

Control. A device designed to regulate the gas, air, water, and electrical supply to a gas appliance. It may be normal or automatic.

Control Clock. A timing device that will at a predetermined time start or stop a control.

Control Cock. A gas valve used in piping to control the gas supply to any section of a system of piping or to an appliance.

Conversion Burner. A burner designed to supply gaseous fuel to an appliance originally designed to use liquid or solid fuel.

Cubic Foot of Gas. The amount of gas that would occupy 1 cubic foot when at a temperature of 60 degrees F., saturated with water vapor and under a pressure equivalent to that of 30 inches of mercury.

Demand. The maximum amount of gas required per unit of time, usually expressed in cubic feet per hour or Btu per hour, required for the operation of the appliance or appliances supplied.

Diversity Factor. Ratio of the maximum probable demand to the maximum possible demand.

Double Oven Combination Gas Range. A domestic range that has an oven and top section using solid or liquid fuel and another oven and section that uses gaseous fuels.

Draft Hood. A device built into an appliance, or made part of the flue or vent connector from an appliance, designed (1) to assure the ready escape of the products of combustion in the event of no draft, back draft, or stoppage beyond the draft hood; (2) prevent a back draft from entering the appliance; and (3) neutralize the effect of stack action of the flue or vent upon the operation of the appliance.

Drip. The container placed at the low point in a system of piping to collect the condensate and from which it may be removed.

Dry Gas. A gas having a moisture and hydrocarbon dewpoint below any normal temperature to which it may be removed.

Dual Oven Combination Gas Range (Dual Fuel, Single Oven). A domestic range designed to operate with either solid or liquid fuel together with gaseous fuels. It has a single oven that may be heated by either the solid or liquid fuel section, or by gas as desired, and a top gas section in addition to the solid or liquid fuel top section.

Duct Furnace. A furnace normally installed in distribution ducts of air-conditioning systems to supply warm air for heating. This definition applies only to an appliance which depends for air circulation on a blower not furnished as part of the furnace.

Excess Air. Air which passes through the combustion chamber and the appliance flues in excess of that which is theoretically required for complete combustion.

Exposed Piping. Gas piping which will be in view in the finished structure.

Flames. (a) Yellow, Luminious, or nonbunsen. The flame produced by burning gas without any premixing of air with the gas. (b) Bunsen. The flame produced by burning gas and air premixed before it reaches the burner ports or point of ignition.

Floor Furnace. A completely self-contained unit furnace suspended from the floor of the space being heated, taking air for combustion from outside this space, and with means for observing flames and lighting the appliance from such space.

Gravity Floor Furnace. A floor furnace depending primarily upon circulation of air by gravity. This classification includes floor furnaces equipped with booster type fans which do not materially restrict free circulation of air by gravity flow when such fans are not in operation.

Fan Floor Furnace. A floor furnace equipped with a fan that provides the primary means for circulation of air.

Flue Collar. That portion of an appliance designed for the attachment of the draft hood or vent connector.

Flue Exhauster. A device installed in and made a part of the vent that will provide a positive induced draft.

Fuel Line. The independent pipe from the meter to an appliance or appliances.

Flue Gases. Products of combustion and excess air.

Flue or Vent. A conduit or passageway, vertical or nearly so, for conveying the flue gases to the outer air.

Furnace (central). A self-contained, gas burning appliance for heating air by transfer of heat from the flue gases through metal to the air and designed primarily to supply heated air through ducts to space remote from the appliance location as well as to the space in which it is located.

Furnace (gravity.) A central furnace that will depend primarily on the circulation of air by gravity. This classification would also include central furnaces equipped with a booster type fan or fans that do not materially restrict free circulation of air by gravity flow when such fans are not in operation.

Furnace (forced-air, central type). A central furnace equipped with a blower that provides the primary means for circulation of air.

Gas Hose. A gas conduit that depends on joint packing for tightness or any wall structure other than formed by one continuous, one-piece metal tubing member.

Gravity. *See* specific gravity.

Heating Valve (total). The number of British thermal units produced by the combustion, at constant pressure, of 1 cubic foot of gas when the products of combustion are cooled to the initial temperature of the gas and air, when the water vapor formed during combustion is condensed, and when all the necessary corrections have been applied.

Hotel and Restaurant Range. A gas appliance of the floor type providing for top cooking, baking, roasting, or broiling, or in any combination of top cooking with any of these other functions and not designed for domestic use.

House Piping. The gas piping beyond the outlet of the meter and extending to the appliance or appliance connector.

House Rise. The principal vertical pipe that conducts the gas from the meter to the different floors of the building.

Incinerator. An appliance used to reduce refuse to ashes, and which is sold as a complete unit.

Portable Incinerator. An incinerator that is a complete unit in itself, and which does not become an integral part of the structure in which it is installed.

Wall Incinerator. An incinerator that is a complete unit in itself designed to be installed in a fireproof wall or chimney, thereby becoming an integral part of the structure in which it is installed.

Individual Main Burner Valve. A valve that controls the gas supply to an individual main burner.

Industrial Gas Boiler. A gas appliance designed primarily to furnish steam for use in some process, the nature of which is industrial or commercial, as distinguished from central heating. This definition would not apply to any boiler covered by standards for space or central heating.

Limit Control. A device responsive to a change in pressure, temperature, or liquid level for turning on, shutting off or throttling the gas supply to an appliance.

Listed Appliance. An appliance that is listed as approved for use as stated by the recognized Testing Agency and recorded in an up-to-date book of registration.

Main Burner. A device or group of devices essentially forming an integral unit for the final conveying of gas and air to the combustion zone and on which combustion takes place to accomplish the function for which the appliance is designed.

Main Burner Control Valve. A valve that controls the gas supply to the main burner manifold.

Manifold. The conduit of an appliance that supplies gas to the individual burners.

Manual Main Shutoff Valve. A valve that controls the gas supply to an appliance, installed in the line, and manually operated to completely turn on or shut off gas to the appliance—excepting the pilots which are provided with independent shutoff valves.

Measured Gas. Gas that has passed through and the volume of which has been registered by a meter.

Meter. The instrument installed to measure the volume of gas delivered through it.

Mixer. The combination of mixer head, mixer throat, and mixer tube.

Mixer Head. That portion of an injection type (Bunsen) type burner, usually enlarged into which primary air flows to mix with the gas stream.

Mixer Throat. That portion of a mixer that has the smallest cross-sectional area and which lies between the mixer head and the mixer tube.

Mixer Tube. That portion of the mixer that lies between the throat and the burner head.

Mixer face. The air inlet end of the mixer head.

Orifice. The opening in a cap, spud, or other device whereby the gas flow is limited and through which the gas is discharged to the burner.

Orifice Cap (hood). A movable fitting having an orifice that permits the flow of gas to be adjusted by changing its position with a fixed needle or other device.

Orifice Spud. A removable plug or cap containing an orifice and which permits adjustment of the flow of gas either by substitution of a spud with a different-sized orifice or by the motion of a needle with respect to it.

Pilot. A small flame used to ignite the gas at the main burner or burners.

Primary Air. The air introduced into a burner and which mixes with the gas before it reaches the port or ports.

Purge. To free a gas conduit of air, gas, or a mixture of gas and air.

Regulator. A device for controlling and maintaining a uniform gas pressure.

Relief Device. A safety device designed to forestall the development of a dangerous condition in the medium of being heated, by relieving pressure, temperature, or vacuum built up in the appliance.

Pressure Relief Valve. An automatic device designed to open or close a relief vent depending on whether the pressure is above or below a predetermined value.

Temperature Relief Valve (fusible plug). A device that opens and keeps open a relief vent by melting or softening of a fusible plug or cartridge at a predetermined temperature.

Reseating or Self-closing Valve. An automatic device that opens and closes a relief vent depending whether the temperature is above or below a predetermined value.

Vacuum Relief Valve. An automatic device that opens or closes a relief vent depending whether the vacuum is above or below a predetermined value.

Relief Opening. The opening provided in a draft hood to permit the ready escape to the atmosphere of the flue products from the draft hood in the event of no draft, back draft, or stoppage beyond the draft hood, and permit the inspiration of air into the draft hood in the event of a strong chimney updraft.

Room Heater. A self-contained free standing non-recessed (except as noted below) gas burning, air heating appliance intended for other than major domestic heating. A room heater may be of the vented or unvented type, except that the unvented room heaters shall not have a normal input rating in excess of 50,000 Btu per hour. Note: this classification also includes unvented, open-flame radiant type wall heaters.

Secondary Air. The air externally supplied to the flame at the point of combustion.

Semirigid Tubing. A gas conduit having a semiflexible wall structure.

Service Pipe. The pipe that brings the gas to the meter from the main gas pipe.

Shutoff. A valve used in piping to control the gas supply to any section of a piping system or to an appliance.

Smoke Pipe. Same as flue or vent connector.

Specific Gravity. As applied to gas specific gravity is the ratio of the weight of a given volume to that of the same volume of air, both measured under the same conditions.

Thermostat. An automatic device actuated by temperature changes, designed to control the gas supply to a burner or burners, in order to maintain temperatures between predetermined limits.

Type A..B..BW..C..Gas vent (see flue or vent).

Type A.. (flue or vent) flues or vents of masonry reinforced concrete, or metal smokestacks.

Special Flues for 1000 F Flue Gas. (Flues for low heat appliances.) A prefabricated listed flue suitable for use with specifically listed gas, oil, wood- or coal-fired low heat appliances where flue temperatures do not exceed 1000 F continuously and do not exceed 1400 F for infrequent brief periods of forced firing.

Type B Gas Flue or Vent. Vent piping of noncombustible, corrosion-resistant material of sufficient thickness, cross-sectional area, and heat insulating quality to avoid excess temperature on adjacent combustible material and listed by a nationally recognized testing agency.

Type BW Gas Flue or Vent. Vent piping of noncombustible material that is corrosion resistant and listed by a nationally recognized testing agency for use only with listed vented recessed heaters.

Type C Gas Flue or Vent. Vent or flue piping of sheet copper of not less than 24 gauge U.S. standard gauge or galvanized iron of not less than 20 U.S. standard gauge or of other approved corrosion-resistant materials.

Unmeasured Gas. Gas that has not passed through and the volume of which has not been measured.

Unit Heater. Low static pressure type. A self-contained, automatically controlled, vented gas-burning appliance, limited to the heating of nonresidential space in which it is installed. Such appliance shall have integral means for circulation of air, normally by a propeller fan or fans and may be equipped with a louvers or face extension made in accordance with the manufacturer's approved specifications.

Unit Heater. High static type. A self-contained, automatically controlled, vented, gas-burning appliance limited to the heating of nonresidential space. These appliances have integral means for installation in the space to be heated unless they are equipped with provisions for attaching both inlet and outlet air ducts.

Vent. *See* flue or vent.

Vent Connector. The flue or vent piping connecting an appliance with the flue or vent. This corresponds to the smoke pipe used with liquid or solid fuels.

Vented Recessed Heaters. A self-contained vented appliance complete with grilles or equivalent designed for incorporation in or permanent attachment to a wall, floor, ceiling or partition, and furnishing heated air circulated by gravity or by a fan directly into the space to be heated through openings in the casing. Such appliances shall not be provided with duct extensions beyond the vertical limits of the casing proper, except that the boots not to exceed 10 inches beyond the horizontal limits of the casing for extension through the walls of nominal thickness may be permitted. Where such boots are provided, they may be supplied by the manufacturer as an integral part of the appliance and tested as such. This definition excludes floor furnaces, unit heaters, and central furnaces as defined elsewhere.

Gravity Vented Recessed Heater. A recessed heater depending on circulation of air by gravity.

Fan Vented Recessed Heater. A recessed heater equipped with a fan.

Wall Heater (unvented, open-flame radiant type). A room heater of the open front type, designed for insertion in or attachment to a wall or partition, having fully exposed flames, and the heat from which is reflected by ceramic radiants or a metal, asbestos, clay brick equivalent back wall reflecting surface. It incorporates no concealed venting arrangements in its construction and discharges all products of combustion through the open front into the room being heated.

Water Heaters. (a) Automatic instantaneous. The type that heats water as it is drawn. (b) Automatic storage. The type that combines a water-heating element and water storage tank, gas to the main burner being controlled by a thermostat. (c) Circulating or tank. Manually controlled type usually connected to the ordinary hot-water tank.

LOCATION OF GAS METERS

Check with the utility supplying gas for regulations on meter locations. Always locate meters where they are accessible for inspection and repair. Keep away from areas that may be hazardous due to electric sparks or unguarded flames. Install where it will not be subject to damage. Where there are corrosive conditions, consult the local utility before locating.

Gas meters have rough-in measurements that will be supplied to the plumber pertaining to headers and large meters upon request to the utility furnishing gas. See information on gas meter illustration.

GREASE TRAPS

Locate grease traps where they are accessible
and at points where grease retention is possible.

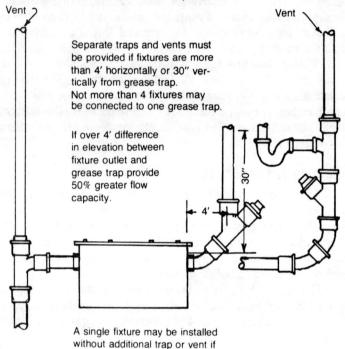

Vent

Vent

Separate traps and vents must
be provided if fixtures are more
than 4' horizontally or 30" vertically from grease trap.
Not more than 4 fixtures may
be connected to one grease trap.

If over 4' difference
in elevation between
fixture outlet and
grease trap provide
50% greater flow
capacity.

30"

4'

A single fixture may be installed
without additional trap or vent if
within 4' or 30" vertically.

Install flow control, check rating
of trap for grease retention and
flow in gallons per minute.

Flow control must be installed and
ample size grease trap provided.

GROUP WASHING EQUIPMENT

Used in factories and places of employment where a number of people want to wash at the same time. This arrangement allows several people at one time to use the same plumbing fixture.

The water tower has a perforated spray head that opens with a foot lever. The water is sprayed in a circle.

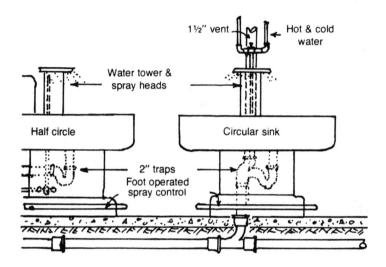

Several types of group washing equipment are in use. Check manufacturer's specifications when roughing in. If P trap under the floor is used, maximum arm to vent must not exceed 5 feet. All horizontal wastes not less than 2 inch pipe. Install cleanout where required. Water piping may rough in above or below the floor. Water is controlled by foot lever bar.

HANGERS AND SUPPORTS

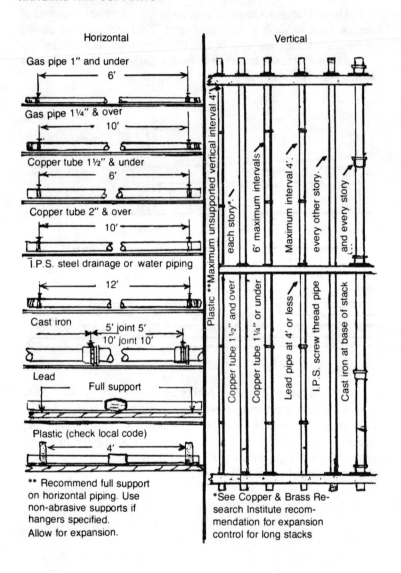

Horizontal

Gas pipe 1" and under
6'

Gas pipe 1¼" & over
10'

Copper tube 1½" & under
6'

Copper tube 2" & over
10'

I.P.S. steel drainage or water piping
12'

Cast iron
5' joint 5'
10' joint 10'

Lead
Full support

Plastic (check local code)
4'

** Recommend full support
on horizontal piping. Use
non-abrasive supports if
hangers specified.
Allow for expansion.

Vertical

Plastic **Maximum unsupported vertical interval 4'

each story*.

6' maximum intervals

Maximum interval 4'.

every other story.

and every story

Copper tube 1½" and over

Copper tube 1¼" or under

Lead pipe at 4' or less

I.P.S. screw thread pipe

Cast iron at base of stack

*See Copper & Brass Re-
search Institute recom-
mendation for expansion
control for long stacks

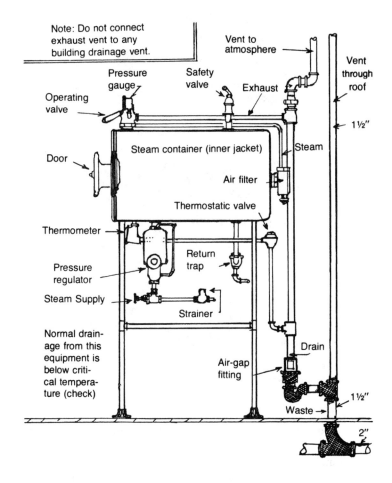

Note: Do not connect exhaust vent to any building drainage vent.

Vent to atmosphere

Vent through roof

Pressure gauge

Safety valve

Exhaust

Operating valve

1½"

Steam container (inner jacket)

Steam

Door

Air filter

Thermostatic valve

Thermometer

Return trap

Pressure regulator

Steam Supply

Strainer

Normal drainage from this equipment is below critical temperature (check)

Drain

Air-gap fitting

Waste

1½"

2"

Hospital equipment such as sterilizers, autoclaves, stills, water-using and other equipment that is in any way connected to the water system or drainage system must be protected by approved air-gaps when entering the building drainage line. Be sure to check for backflow prevention devices on any water-connected hospital equipment.

BED-PAN WASHERS AND EQUIPMENT

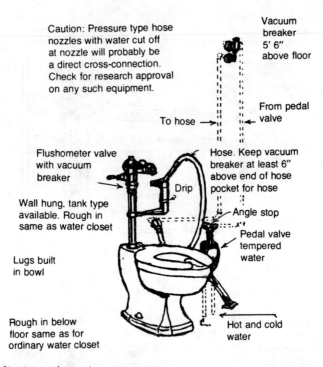

Caution: Pressure type hose nozzles with water cut off at nozzle will probably be a direct cross-connection. Check for research approval on any such equipment.

Vacuum breaker 5' 6" above floor

From pedal valve

To hose →

Flushometer valve with vacuum breaker

Wall hung, tank type available. Rough in same as water closet

Lugs built in bowl

Rough in below floor same as for ordinary water closet

Drip

Hose. Keep vacuum breaker at least 6" above end of hose pocket for hose

Angle stop

Pedal valve tempered water

Hot and cold water

Check manufacturer's measurements

Note: Vacuum breakers serving such equipment must be installed at least 6 inches above high point of usage. Long hoses may constitute a hazard when bracket for support is installed too high.

Check for approved vacuum breakers and backflow devices on all hospital equipment. Vacuum breaker must be 36 inches above fixture and 5 feet 6 inches above floor.

HOSPITAL EQUIPMENT

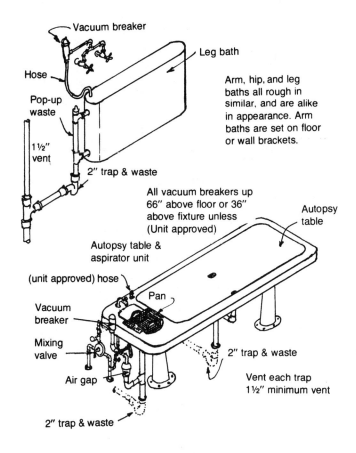

All hospital equipment must be installed with proper anti-siphon devices, air-gaps, and vacuum breakers. Check with local codes and install to manufacturers measurements and specifications.

WATER HEATERS (180 & 140 DEGREE)

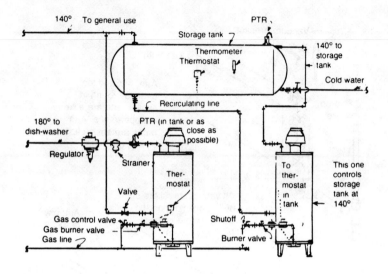

FORCED HOT WATER HEATING SYSTEM

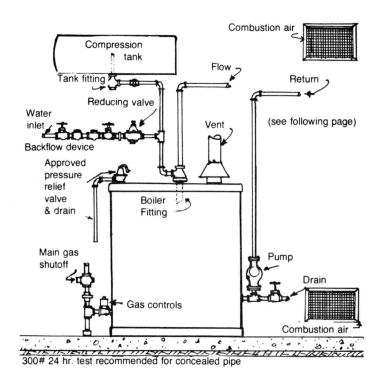

Combustion air

Compression tank

Flow

Return

Tank fitting

Reducing valve

Water inlet

Vent

(see following page)

Backflow device

Approved pressure relief valve & drain

Boiler Fitting

Main gas shutoff

Pump

Drain

Gas controls

Combustion air

300# 24 hr. test recommended for concealed pipe

Note: Piping between boiler fitting and tank fitting must be a minimum of ¾ inches in size. Horizontal sections of this pipe must be pitched upward at least ¼ inch per foot. Backflow prevention device may be required by some jurisdictions (check). Use only approved pressure relief values. Drain to point as specified by code. Termination of drain should not be threaded. Install all hot water heating systems strictly according to approved installation instructions.

Properly installed hot water heating systems will provide a safe, economical, and uniform distribution of heat throughout the house. It is virtually dust free, reduces noise, decorating, and maintenance costs.

Check the following points. Test as required.

1. Provide heat loss calculations and size pipe and heater accordingly.

2. Install according to manufacturer's installation instructions and local code requirements. Check for backflow prevention device.

3. Provide combustion air top and bottom within 6 inches of floor and ceiling. Minimum air in equal parts at the ratio of 100 square inches plus 1 square inch per 1000 Btu gas appliance input rating.

4. Install approved ASME relief valve and drain. Be sure the relief valve is sized and rated by a nationally recognized testing agency for the capacity of the heater.

5. Check full-size drain point outside of house within 2 feet of ground and not closer than 6 inches. Do not thread end of drain. Do not use soft copper or pipe that may be bent, broken, or damaged. Hard copper or steel pipe is recommended (check with local inspector).

6. Check for approved gas shutoff in accessible position.

7. Check nameplate for type of gas and clearances.

8. Provide proper clearance front, sides, and rear of appliance. Provide for 30 inches working space in front.

9. Check gas pipe for adequate supply.

10. Install proper size vent connector to flue. (No single wall metal flue within 6 inches of combustible materials.)

11. Check location of vent termination. (Keep 1 foot above and 4 feet away from house openings.)

12. Support compression tank and all piping.

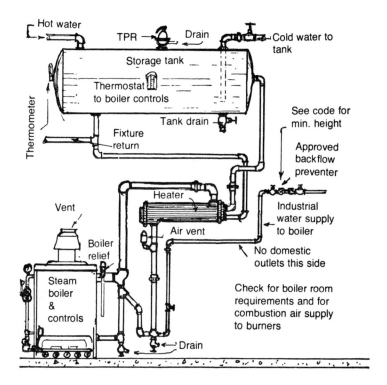

Install pressure relief valves where there is a possibility of a closed system. All boiler and storage tank relief valves should be ASME approved; rating and discharge capacity.

All water supply to stream boiler must be protected by an approved backflow preventer or be installed to a nonpotable supply source.

All steam boiler installations must be installed according to safety rules and requirements. Check for rules on boiler rooms and combustion.

INTERCEPTORS

All built-on the-job interceptors must be designed to properly screen all wastes. All wastes that contain solid matter that may cause a stoppage of the public sewer must be removed before entering the sewer. Each waste to be treated for solids by the method required by the department having jurisdiction. Wastes may be by gravity or may be pumped to the screening section of the interceptor. Provide sump with deep seal trap as settling tank before entering sewer. Manufactured interceptors may be used when possible.

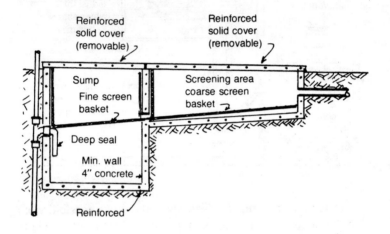

All concrete tanks must be properly engineered to ensure a leakproof, gas-tight tank and to withstand all loads on tanks or covers. Cement plaster inside.

All sumps or settling tanks must be properly trapped and vented; no cleanouts inside of sump. The use of ladders or removal of equipment to service interceptors is prohibited. Do not use interceptor as a collection tank for any other wastes.

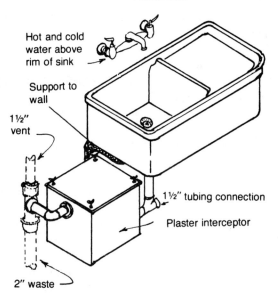

Plaster sink

Hot and cold water above rim of sink

Support to wall

1½" vent

1½" tubing connection

Plaster interceptor

2" waste

A plaster sink is usually installed where there is a need to mix plaster for plaster casts and similar use.

An interceptor that acts as a settling basin is installed adjacent to the sink, in an accessible position and is fitted with a cover that is easily removed for frequent cleaning of the receptor.

Plaster receptors are required where wastes from mixing are allowed to enter the sewer or drainage system.

INTERCEPTORS INDUSTRIAL

All built-on-the-job industrial sand traps, wash racks, or other required interceptors must be designed to adequately perform their function.

1. Determine capacity load to be treated.

2. Design concrete for 3000 psi. Reinforce all walls and covers to withstand anticipated load.

3. Elevate above grade if outside. No surface rainwater to enter waste drainage system.

4. Locate in an accessible place. All covers to be readily removable.

5. Locate grate on inlet compartment only.

6. Provide 18 inches square "sample box" with easily removable cover.

7. Minimum seal for traps 6 inches. (Check for maximum.)

8. Waterways may be "slotted" in concrete or pipe cast into walls. Pipe size to equal inlet.

9. No cleanout serving drainage piping inside box.

10. Inlet to be full pipe diameter above outlet.

11. Vent as required by code.

12. When airtight closed covers are used, design to prevent air lock.

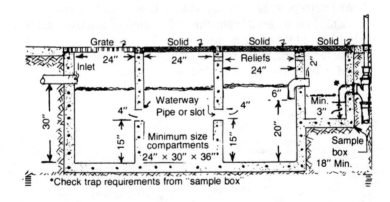

*Check trap requirements from "sample box"

Industrial interceptors are required when wastes with heavy suspended solids or wastes with grease or oil must enter the sanitary system. Check all cannery, food processing, poultry or meat packing plants, equipment washing, laundry and cleaning plants, factories, etc. See previous two pages.

Details shown below may vary. Size will depend on amount of waste to be treated. For tanks that do not require engineering, the detail will meet most code requirements.

1. Inlets may require screens below. Grated top may be used on inlet side only.

2. Covers must be readily removable. Provide handholes or other means.

3. Elevate if on outside. If located on outside, design to prevent rain or surface water from entering. Large-scale wash racks must be under roofs. Provide curbs if necessary.

4. No domestic wastes permitted. Each interceptor to be designed for industrial use only.

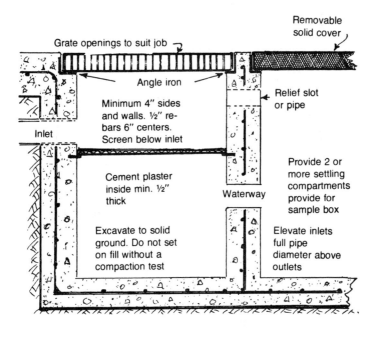

Removable solid cover

Grate openings to suit job

Angle iron

Minimum 4″ sides and walls. ½″ re-bars 6″ centers. Screen below inlet

Inlet

Cement plaster inside min. ½″ thick

Excavate to solid ground. Do not set on fill without a compaction test

Relief slot or pipe

Waterway

Provide 2 or more settling compartments provide for sample box

Elevate inlets full pipe diameter above outlets

VENTILATING AN INTERIOR BATHROOM

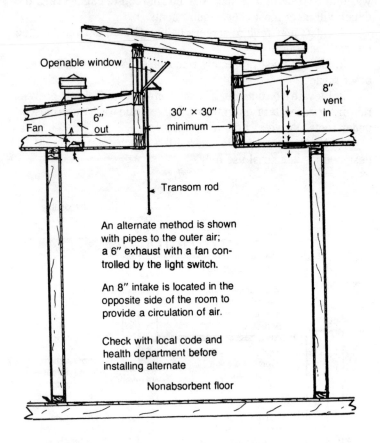

Openable window

Fan

6" out

30" × 30" minimum

8" vent in

Transom rod

An alternate method is shown with pipes to the outer air; a 6" exhaust with a fan controlled by the light switch.

An 8" intake is located in the opposite side of the room to provide a circulation of air.

Check with local code and health department before installing alternate

Nonabsorbent floor

Two ways to ventilate an interior bath are shown. The number one choice is the ventilated skylight with an openable window. The minimum size for dwelling units is 3 square feet. For hotels and apartments it is 6 square feet.

ISLAND SINK INSTALLATION

Check with the local inspection agency

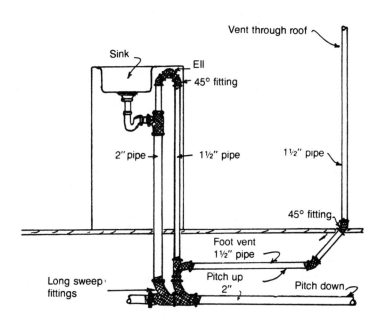

One method of installing an island sink. Raise return vent as high as possible under the drain board. Use two (45-degree) fittings and one long radius elbow to make loop. Install long sweep fitting below the floor and run horizontally to nearest partition. Use 45-degree fittings for easy stoppage cleaning.

JOINTS AND CONNECTIONS

Screw Joints. These joints join water, gas, and drainage pipe together by an American Standard thread that enters the receiving fitting and is tightened with a wrench. Normally, a compound called pipe dope is applied to the outer thread as a filler to make a tight joint. Never apply dope to the thread of the fitting becauuse the dope will usually work back into the pipe and so into the water supply of the house.

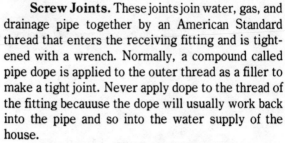

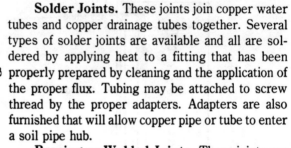

Solder Joints. These joints join copper water tubes and copper drainage tubes together. Several types of solder joints are available and all are soldered by applying heat to a fitting that has been properly prepared by cleaning and the application of the proper flux. Tubing may be attached to screw thread by the proper adapters. Adapters are also furnished that will allow copper pipe or tube to enter a soil pipe hub.

Brazing or Welded Joints. These joints are specified when extra strength is desired. Brazing and welding are accomplished by fusing the proper metal rod to the pipe by means of extreme heat. This is a specialized trade and should be only done by proficient mechanics as it is possible to build a joint that may look and even test all right, but may fail because of poor fusing.

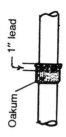

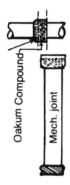

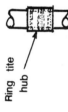

Lead and Oakum. A joint used in connecting cast-iron hub and spigot pipe. This joint is used also in connecting any approved drainage to a cast-iron hub. The joint is made by compacting oakum with the proper tools to within 1 inch of the top of the hub and then pouring a ring of lead of not less than 1 full inch in depth to fill the joint. To finish the joint and make a leakproof connection, caulking is required.

Clay Pipe. Clay pipe may be joined by a hot poured compound or mechanical joints as approved by code. All hot-poured joining compounds must meet a certain standard for melting and for adhesive qualities. Mechanical joints are as approved for and only approved mechanical joints should be used. Check with your local inspection agency before installing sewers, for approved joints. An approved joint is shown at left. This one has a molded rubber-like compound that is fused to the clay within the hub and on the spigot end. To join simply lubricate and slip together.

Cement Asbestos. This material has a patented joint that slips together with a lubricant. All such pipe and fittings must be marked for sewers as this material is used extensively as gas venting pipe and must not be confused. When using cement asbestos materials, consult your local inspection department for specifications on installation for this and its connection to other materials.

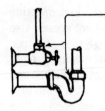

Slip Joints. These joints slip into a milled socket that is threaded on the outside. The slip-joint nut squeezes a gasket or packing by the action of tightening the nut to the thread of the milled socket. These joints are prohibited where they are concealed in any way. Gasket type slip joints are prohibited on the sewer side of the trap on drainage.

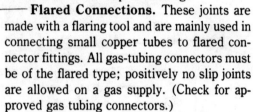

Flared Connections. These joints are made with a flaring tool and are mainly used in connecting small copper tubes to flared connector fittings. All gas-tubing connectors must be of the flared type; positively no slip joints are allowed on a gas supply. (Check for approved gas tubing connectors.)

Unions. These fittings join pipes together that meet from different starting points. Many types are on the market; the most common is the ground joint union and the flange union.

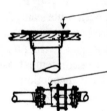

Flanged Fixture Connection. This is a joint that is made bolting a fixture to an especially installed floor or wall flange. A toilet connection is a flanged fixture connection.

Expansion Joint. This joint is installed on lines that are subject to movement due to expansion or contraction.

124

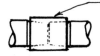

Bituminous Fiber. This joint has a tapered end that fits a matching tapered hub and is driven together with any driving tool. No lubricant is necessary as it fits tight and is a milled joint. The material is a wood fiber mixed with patented additives, dried, and impregnated with tar under extreme pressure.

Ceramic Weld. This joint has an inner rubber hub encased by an outer hub. To fit the pipe securely, bands of stainless steel cinch the rubber tightly to the pipe by the action of tightening bolts attached to the bands with a socket wrench. After the joint has been made and tested, the outer casing is filled with soft grout to complete the joint and to ensure a permanent hub.

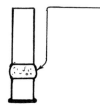

Wiped Joints. These joints join lead to lead or lead to brass. The method is an art that must be learned and consists of applying a molten mixture of lead and tin to the pipe or the pipe and ferrule by spashing on with a paddle or pouring on with a ladle. The shape and thickness of the joint is controlled by wiping cloths held in the hand of the mechanic. Lead working is a specialized trade and many present-day plumbers have no knowledge of the science.

Burned Joints. This joint is made by welding lead together by means of fusing lead to lead.

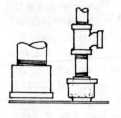

Air-gap fittings for indirect wastes. Eliminates floor sinks and provides a rigid connection at sewer entry.

These fittings are inventions of the author and are "code" approved.

Tapped hub adapters for joining threaded pipe to cast-iron hubs.

This fitting is a soil pipe bushing with approved recessed thread and designed smooth inside waterway. Permits an approved economical method of connection.

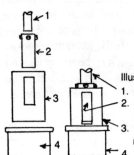

Illustrated is the Smith air gap.
1. Drain from fixture, into . . .
2. Splash tube with centering screws, into . . .
3. Barrel of air gap, slips into receiving tapped hub adapter
4. Tapped hub adapter.

NO-HUB, CAST-IRON JOINTS

A new development by the cast-iron industry for eliminating lead and oakum joints. A metal covering is tightened by steel bands compressing the gasket.

GLASS PIPE JOINTS

A heavy-metal outer joint tightens and centers an interior washer type gasket. Glass pipe is usually flared on each end, and the joint lined up and clamped tight.

BUILDING SEWER BAND SEAL JOINT

A compression joint tightened by metal bands similar to hose connections. This is similar also to the ceramic weld coupling and the cast-iron joint except it has no outer covering.

126

(ABS)   (PVC)

Plastic Pipe and Fittings. Recent introduction into the housing field, these materials are generally required to be solvent welded together by painting the specified cementing material to the inner socket and the exterior of the pipe equal to the socket length. Excessive solvent is to be avoided. All surfaces to be joined must be completely free of all foreign matter. Upon preparing the joint, force together immediately with a slight twisting motion to ensure full engagement of pipe to end of socket.

(ABS) Acrylonitrile-Butadiene-Styrene.

(PVC) Polyvinyl-Chloride.

The following rules for a proper installation are as follows:

1. Install strictly in accordance with manufacturer's instructions and local code requirement.

2. Follow safety orders in handling and storage.

3. Check use of material. Many plastics will not stand up under certain solvents, thinners, or chemicals.

4. Provide for expansion and contraction. This must be checked as plastics vary in length as temperatures change. Be sure to check manufacturer's expansion table.

5. Check supports. Do not install supports or hangers that could have an abrasive action on the surface of the pipe. Pipes passing through walls or timbers should be protected against the possibility of nail penetration.

6. Check for approved material and proper identification. Use approved adapters to other pipe.

7. Check for prohibited locations. Some plastics burn readily and will transfer flame, smoke, and fumes from one point to another. See requirements for exterior exposure of material used.

LAWN SPRINKLER SYSTEMS

Vacuum breakers
are required on
lawn sprinkler
systems to prevent
a possible contamination
of the domestic water supply
by back siphonage.

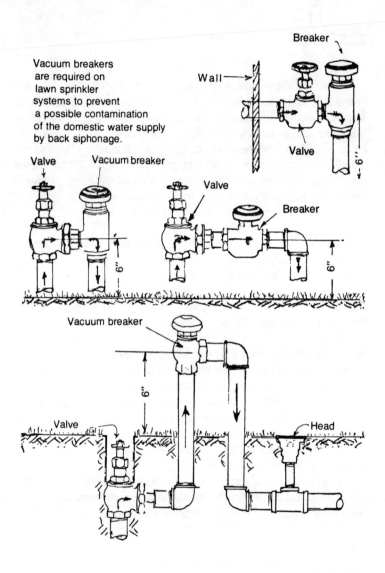

LADDERS CONSTRUCTED ON THE JOB

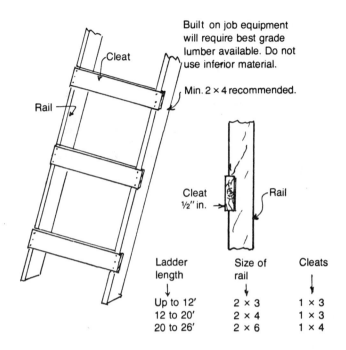

Built on job equipment will require best grade lumber available. Do not use inferior material.

Min. 2 × 4 recommended.

Cleat

Rail

Cleat ½" in.

Rail

Ladder length	Size of rail	Cleats
Up to 12'	2 × 3	1 × 3
12 to 20'	2 × 4	1 × 3
20 to 26'	2 × 6	1 × 4

Minimum nailing for cleats are 3 10d for each side. All cleats are gained into rail ½ inch or are blocked between. Ladders shall be secured against displacement by:

1. Nailing or lashing the ladder at the top.
2. Nailing cleats on floor in front of ladder.
3. Means that will prevent any displacement.

Ladders serving platforms shall extend 2 feet to 6 feet above upper landing. Note: All state safety orders must be complied with. Refuse to work with equipment that may be dangerous in any way. Check equipment before use.

129

LIQUEFIED PETROLEUM GAS (LPG)

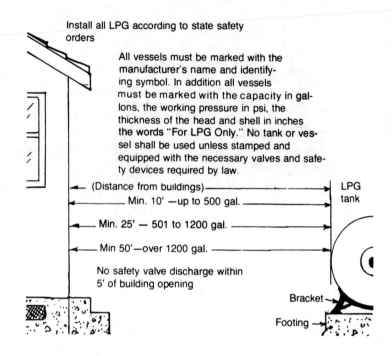

Install all LPG according to state safety orders

All vessels must be marked with the manufacturer's name and identifying symbol. In addition all vessels must be marked with the capacity in gallons, the working pressure in psi, the thickness of the head and shell in inches the words "For LPG Only." No tank or vessel shall be used unless stamped and equipped with the necessary valves and safety devices required by law.

(Distance from buildings)
Min. 10' —up to 500 gal.
Min. 25' — 501 to 1200 gal.
Min 50'—over 1200 gal.

No safety valve discharge within 5' of building opening

LPG tank

Bracket
Footing

All LPG house piping must conform to state and local safety orders. All piping and valves must be capable of withstanding the pressures required by the safety orders. All pipe dope used must be insoluble in LPG or water. Bury pipe 2 feet deep and protect with corrosion resistant material. Bushings, street ells, and street tees are prohibited. Test all piping at 20 pounds psi or better. All appliances must be labeled for the type of gas used.

PIPE CAPACITORS (Soil and Wastes)

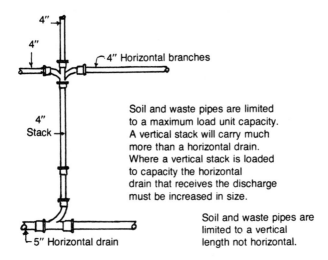

4″

4″

4″ Horizontal branches

4″
Stack

Soil and waste pipes are limited to a maximum load unit capacity. A vertical stack will carry much more than a horizontal drain. Where a vertical stack is loaded to capacity the horizontal drain that receives the discharge must be increased in size.

Soil and waste pipes are limited to a vertical length not horizontal.

5″ Horizontal drain

PIPE CAPACITIES (Vents)

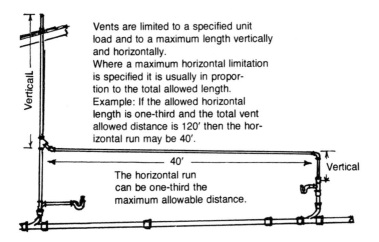

Verticall

Vents are limited to a specified unit load and to a maximum length vertically and horizontally.
Where a maximum horizontal limitation is specified it is usually in proportion to the total allowed length.
Example: If the allowed horizontal length is one-third and the total vent allowed distance is 120′ then the horizontal run may be 40′.

40′

Vertical

The horizontal run can be one-third the maximum allowable distance.

131

GALLONS DISCHARGED PER MINUTE BY SEWER PIPE

Size	Fall Per 100 Feet							
	1" Fall	2" Fall	3" Fall	6" Fall	9" Fall	12" Fall	24" Fall	36" Fall
3"	9	12	15	22	27	31	44	54
4"	20	28	35	50	62	71	101	124
6"	62	89	111	156	194	224	317	389
8"	140	198	246	348	432	499	706	864
9"	196	277	339	480	595	687	971	1180
10"	261	369	457	648	803	928	1310	1180
12"	432	612	758	1070	1330	1530	2170	2660
15"	800	1130	1400	1980	2450	2830	4010	4910
18"	1320	1860	2310	3260	4040	4660	6590	8080
20"	1720	2500	3060	4330	5305	6130	8660	10610
24"	2010	4110	5035	7191	8810	10270	14520	17790
27"	4020	5680	6960	9840	12050	13920	19680	24110
30"	5380	7618	9320	13180	16140	18640	26350	32280
33"	6950	9840	12050	17040	20865	24090	34070	41730
36"	8800	12450	15210	21565	26410	30500	43130	52820

(For old cast iron drains see page 248)
To read chart: Follow size of pipe across to where the pitch or
fall is shown that corresponds with the known pitch of the sewer.
12" fall = ⅛" per ft. or 1%   24" fall = ¼" per ft. or 2%

To read chart, follow size of pipe across to where the pitch or fall is shown that corresponds with the known pitch of the sewer: 12 inches fall = ⅛ inch per foot or 1%. ... 24 inches fall = ¼ inch per foot of 2%.

SIZING OF DRAINS AND VENTS

All wastes and vents are assigned certain unit values to determine capacities of these pipes. Vent values are as follows: (Combined vertical and horizontal) *Maximum horizontal, ⅓ of combined.

Size - - - - - 1¼"	1½"	2"	2½"	3"	4"	5"	6"
Units	1	8	24	48	84	256	600	1380
Total Length in Ft.	45'	60'	120'	180'	212'	300'	390'	510'
* Horz. Max. Ft.	0	20'	40'	60'	71'	100'	130'	170'

No. 6 unit trap be connected to less than a 2-inch vent pipe. No. 1¼" inches horizontal vent allowed.

Drainage values are as follows: (Check 3" limits)							
Size - - - - 1¼"	1½"	2"	2½"	3"	4"	5"	6"
Max. Units Vert.	0	2	16	32	48	256	600	1380
Max. Units Horiz.	0	1	8	14	35	180'	356	600
Max. Length Vert. Et.	0	65'	85'	148'	212'	300'	390'	510'

There is no limitation on the horizontal length. Only grade or fall would limit the distance that a gravity drain can be run.

3 and 4 inches, traps are rated as 6 units (includes 50 gpm).

2 inches, traps are rated as 3 units (up to 30 gpm).

1½ inches, traps are rated as 2 units (up to 14 gpm).

1¼ inches, traps are rated as 1 unit (up to 7½ gpm).

Check for special unit ratings for urinals and other special fixtures requiring larger ratings.

TYPICAL 3-INCH HOUSE DRAIN

A full-size vent stack is required in almost any building. Exceptions are where the aggregate of all the vents equals or exceeds the main vent or where a full-size vent may be reduced in size. No less than a 2-inch vent may serve a toilet or similar fixture that discharges or is rated as equal.

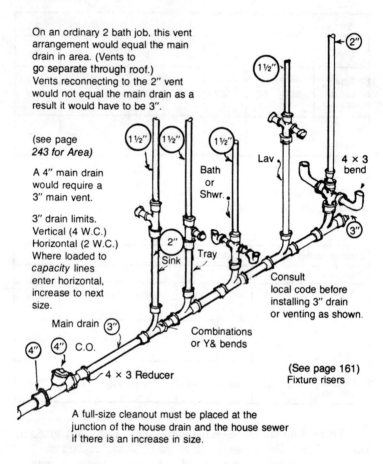

On an ordinary 2 bath job, this vent arrangement would equal the main drain in area. (Vents to go separate through roof.) Vents reconnecting to the 2″ vent would not equal the main drain as a result it would have to be 3″.

(see page *243 for Area*)

A 4″ main drain would require a 3″ main vent.

3″ drain limits. Vertical (4 W.C.) Horizontal (2 W.C.) Where loaded to *capacity* lines enter horizontal, increase to next size.

Main drain

C.O.

4 × 3 Reducer

Bath or Shwr.

Sink

Tray

Lav

4 × 3 bend

Combinations or Y& bends

Consult local code before installing 3″ drain or venting as shown.

(See page 161) Fixture risers

A full-size cleanout must be placed at the junction of the house drain and the house sewer if there is an increase in size.

PIPE CAPACITIES IN WATER-SUPPLY PIPE

New Fairly Smooth Pipe			Estimated Flow Percentage Various Pipe Conditions		
Pipe Size	Actual Diameter	Flow (gpm)	(1)	(2)	(3)
⅜"	0.488	2.45	71.4	39.6	10.6
½"	0.618	4.55	72.6	45.4	18.5
¾"	0.820	9.60	75.0	47.4	27.1
1"	1.04	18.0	75.6	48.6	32.5
1¼"	1.37	37.6	75.8	50.7	36.4
1½"	1.60	56.0	75.9	53.2	40.4
2"	2.06	108.0	76.4	54.6	44.0
2½"	2.46	173.0	76.6	54.7	45.7
3"	3.06	310	76.8	54.8	47.4
4"	4.02	630	77.0	54.8	49.2
5"	5.04	1125	77.7	55.0	51.1
6"	6.05	1805	78.6	55.2	51.5

The tables above are for ordinary water piping conditions standard steel or iron pipe. The flow is figured for a friction loss of approximately 10 pounds/inches per 100 feet of pipe.

When sizing water pipe for use over a long period of time, conditions for service must be considered such as possible corrosion and caking that will restrict the flow as time goes on.

Condition 1. This condition may be used for an estimated future condition on which the pipe will be corroded, slightly caked, or fairly rough.

Condition 2. This condition may be used for badly caked, very rough, and corroded condition.

Condition 3. This use is for a very badly caked, very rough, and corroded condition.

New fairly smooth pipe will give the flow for a new installation. To find the ultimate demand, multiply the percentage for the estimated use.

Plumbing Symbols			
Item	**Plan**	**Letter**	**Symbol**
Drainage line	◯	D.	
Vent line	◯	V.S.	
Tile pipe	◯	V.C.P.	
Cold water line	◯	C.W.	
Hot water line	◯	H.W.	
Hot water return	◯	H.W.R.	
Gas pipe	⊗	G.	
Ice water supply	◯	D.W.	
Ice water return	◯	D.R.	
Fire line	◯	F.L.	
Indirect waste	◯	I.W.	
Industrial Sewer	⊕	I.S.	
Acid waste	◯	A.W.	
Air line	Ⓐ	A	
Vacuum line	Ⓥ	V	
Refrigerator waste	Ⓡ	R	
Gate valves			
Check valves			
Cleanout		CO.	
Floor drain		F.D.	
Roof drain		R.D.	
Refrigerator drain		REF.	
Shower drain		S.D.	
Grease trap		G.T.	
Sill cock		S.C.	
Gas outlet		G	
Vacuum outlet		Vac.	
Meter		M	
Hydrant			
Hose rack		HR	
Hose rack built in		H.R.	
Leader		L	
Hot water tank		H.W.T.	
Water heater		W.H.	
Washing machine		W.M.	
Range boiler		R.B.	

Pipe color code: city, water, white; air, grey; sanitation drains and vents, blue; fire lines, red; acid, industrial water, indirect waste, green; gas, brown orange; vacuum, cream.

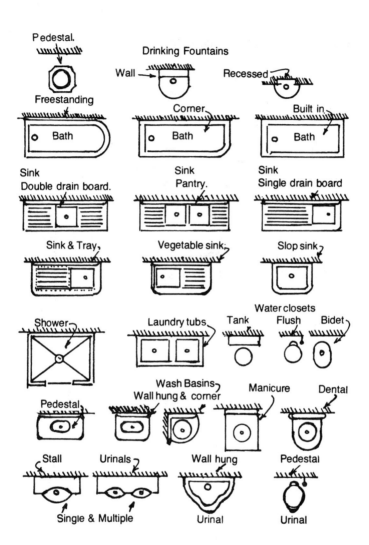

Pedestal.

Drinking Fountains

Wall →

Recessed

Freestanding

Bath

Corner

Bath

Built in

Bath

Sink
Double drain board.

Sink
Pantry.

Sink
Single drain board

Sink & Tray

Vegetable sink.

Slop sink

Shower

Laundry tubs

Water closets

Tank   Flush   Bidet

Pedestal

Wash Basins
Wall hung & corner

Manicure

Dental

Stall

Urinals

Wall hung

Pedestal

Single & Multiple

Urinal

Urinal

137

PROHIBITED FITTINGS AND CONNECTIONS ON DRAINAGE PIPING

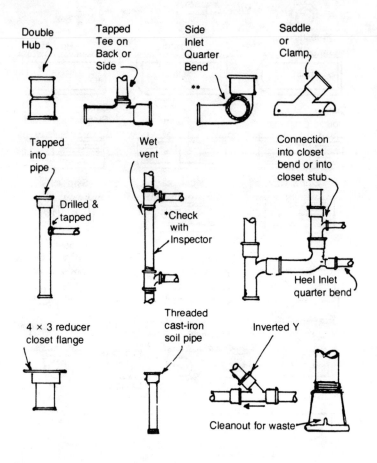

Double Hub

Tapped Tee on Back or Side

Side Inlet Quarter Bend **

Saddle or Clamp

Tapped into pipe — Drilled & tapped

Wet vent — *Check with Inspector

Connection into closet bend or into closet stub

Heel Inlet quarter bend

4 × 3 reducer closet flange

Threaded cast-iron soil pipe

Inverted Y

Cleanout for waste

Double hubs, inverted Y's, and side inlet quarter bends may be used for vents.

*Where wet vents allowed see limitations.

**Vent in vertical position only.

PROTECTION OF MATERIALS

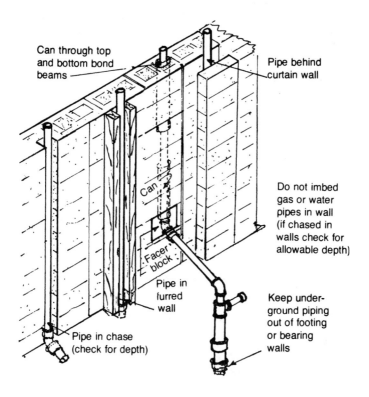

Can through top and bottom bond beams

Pipe behind curtain wall

Can

Do not imbed gas or water pipes in wall (if chased in walls check for allowable depth)

Facer block

Pipe in furred wall

Keep underground piping out of footing or bearing walls

Pipe in chase (check for depth)

Pipe must not be imbedded in concrete walls or footings. All piping in such walls must be protected against settlement, expansion, contraction, and corrosion.

Where allowed, piping may enter hollow-core walls provided cans are installed at bond-beam locations to allow piping to run through. Where pipes enter hollow walls, facer blocks should be provided to allow for tests and inspection.

All piping passing through concrete footings or walls must be protected against settling, expansion, contraction, and corrosive action. No piping may be built in or imbedded in concrete or masonry walls or footings.

No plumbing should cross through walls or footings unless adequately protected from damage by proper sleeves or cans at least ½ inch larger than the piping.

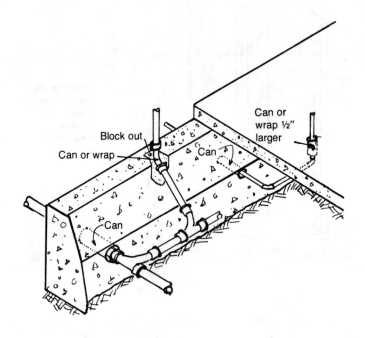

Pipes passing through blockouts should be thoroughly protected by canning or wrapping as approved by the local inspection agency. Piping under concrete should be protected against corrosion as directed by local code. Gas piping under concrete slabs is prohibited by most inspection departments; if allowed it will have restrictive provisions. Check for piping regulations in or under concrete construction.

POTATO PEELER INSTALLATION

Potato peelers and similar food processing appliances are installed with an indirect connection to the sewer. Normally a floor receptacle is roughed in directly below the drain of the appliance.

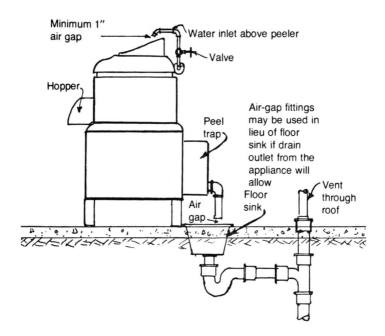

Receptor and waste must be large enough to carry away all drainage without flooding. Do not hook up to grease trap. Peel trap required to catch all peelings and bits of matter that may cause stoppages.

If the potato or vegetable peeler is so constructed that the appliance has a built in air gap, a direct hookup may be allowed. Check with the local inspection department on food appliance installations when in doubt.

RANGE HOOD GRAVITY TYPE

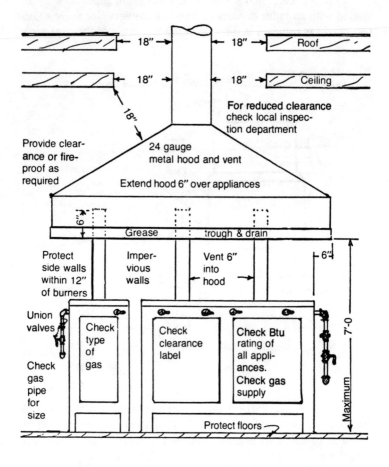

All walls behind hood and appliances to be fireproofed to meet local requirements. Size gas pipe for combined demand of all appliances. Check with your fire marshall for state laws.

RANGE HOOD WITH MECHANICAL VENTILATION

Exhaust capacity: Exhaust duct systems must create a minimum air velocity of not less than 1500 feet per minute and not more than 2200 feet per minute. Replace air exhausted by equipment.

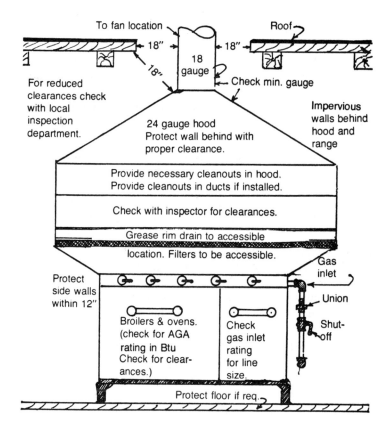

Check with your fire marshall for state requirements. Appliances without stamped clearances allow 18 inches clearance sides, top, bottom, and rear. Note: For inside duct, provide continuous shaft fireproofed as required by building regulations.

RESTAURANT RANGE INSTALLATION

1. Check range for approval label and burner rating, type of gas and pipe size inlet.

2. Pipe size. Determine length of pipe from supply to range and size accordingly.

3. Clearances. check approval label plate for clearances to combustibles. Check vent from canopy through roof. Check vent clearances required to combustibles. Check floor under range for protection if set on combustible floors. 18 inches clearance required to all combustible materials if canopy smoke pipe is single wall metal. Check wall construction behind canopy, protect as required by local ordinance.

4. Gas shutoff. Install a gas cock to control range. Each separate appliance is to have separate control. Approved plug valve required by most inspection departments.

5. Unions. Install union between valve and appliance. Approved ground joint only in accessible position.

6. Size of canopy. Canopy must extend at least 6 inches beyond any appliance or group of appliances it is located under. Walls and floors must be of nonabsorbent materials that is impervious to grease and that may be easily cleaned. Canopy must have a grease collection through that has an accessible drain.

7. Vents from appliances. All vents must extend up into the canopy at least 6 inches. Vents that are constructed to serve as a manifold must be of adequate size to serve as such. Water heater vents should not vent into the canopy but should be entirely independent of all other appliance vents.

8. Do not install a restaurant range to a supply gas line until it is determined that the Btu rating of the appliance and the cubic feet per hour required will supply the necessary amount of gas. Do not install LPG gas to a burner that is designed for natural gas.

9. Consult with local fire department for laws effecting restaurant installations.

RAINWATER DRAINAGE INSTALLATION

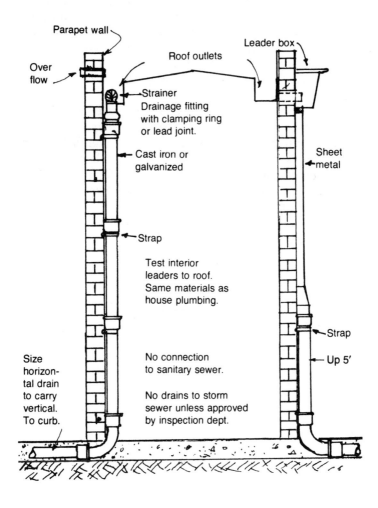

Parapet wall

Leader box

Over flow

Roof outlets

Strainer
Drainage fitting
with clamping ring
or lead joint.

Cast iron or
galvanized

Sheet
metal

Strap

Test interior
leaders to roof.
Same materials as
house plumbing.

Strap

Up 5'

Size
horizon-
tal drain
to carry
vertical.
To curb.

No connection
to sanitary sewer.

No drains to storm
sewer unless approved
by inspection dept.

145

RAIN LEADER INSTALLATION

Interior leaders are installed with the same materials allowed for house drains. All inside leaders are tested to the roof with water to ensure leakproof installation. No sheet-metal leader is allowed for an interior installation. Exterior or outside leaders may be installed with sheet-metal down spouts of weights allowed by code. Piping underground may be of any materials allowed for underground drainage.

Where outside down spouts abutt a public street or sidewalk, install approved house drainage materials 5 feet up from finished grade and strap securely. When installing piping for underground leaders, run cast iron or equal 2 feet from building line before installing clay, cement asbestos, or bitumious fiber pipe. Do not connect any leader pipe to the sanitary sewer unless approved by the sewer district engineer having jurisdiction.

Following is a table of pipe sizes and drainage areas (Vertical leaders to surface drainage).

Pipe Size	Drainage Area Sq. Ft.
1¼″	360
1½″	500
2″	1,000
2½″	1,500
3″	2,460
4″	5,220
5″	9,430
6″	15,030
8″	31,490
10″	55,150
12″	88,250

The table above has been used successfully for vertical leaders in central California. Where rainfall is heavy the drainage area may be less. When vertical leaders enter horizontal storm drains, increase to the next larger pipe. Size the drain according to local rainfall conditions. Do not enter sanitary sewers unless approved. Screen all roof-surface drains before entering leaders.

USE OF ROPES

Rope is used frequently in the plumbing of a building as a means of hoisting up sections of piping and equipment preparatory to hanging or fastening. Care should be used in selecting a rope that will be completely safe. Following is a safe load capacity for good manila rope. Be sure to use the proper knot for intended use.

Diameter in Inches	Circumference in Inches	Weight per 100 Feet in Pounds	Safe Load Capacity in Pounds
¼	¾	1.71	120
⅜	1	3.45	260
½	1½	7.36	520
⅝	2	13.10	880
¾	2¼	16.40	1080
⅞	2¾	22.00	1300
1	3	26.50	1800
1⅛	3½	35.20	2400
1¼	3¾	40.80	2700
1½	4½	58.80	3700
1¾	5½	87.70	5300
2	6	105.00	6200
2½	7½	163.00	9300
3	9	273.00	13800

Consult Federal Standards for rope requirements.

Inspection of Rope. Before using rope that will be used for heavy or dangerous work, check it over thoroughly by laying open the strands. If the inner fibers of the strands are broken or frayed or the rope has a mildewed appearance or a musty odor, it might be unsafe.

When lifting heavy objects, use guidelines and keep workmen in the clear. Secure all work with the proper hangers or fasteners as soon as possible after lifting.

USEFUL KNOTS

A Bowline knot is one of the most commonly used knots. Properly tied it will not slip.

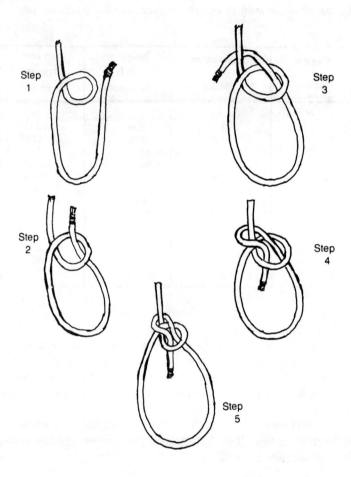

Step 1

Step 2

Step 3

Step 4

Step 5

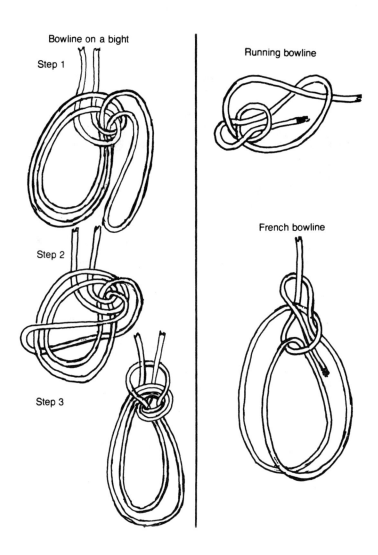

Bowline on a bight

Step 1

Step 2

Step 3

Running bowline

French bowline

149

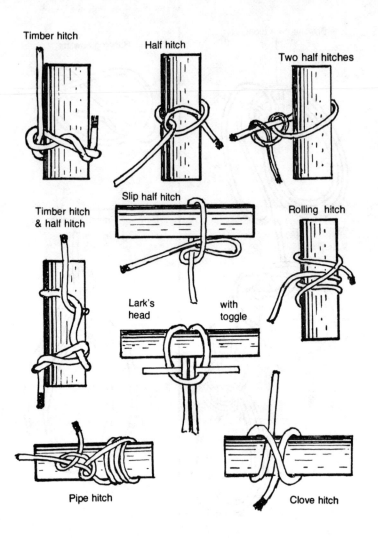

Timber hitch

Half hitch

Two half hitches

Timber hitch & half hitch

Slip half hitch

Rolling hitch

Lark's head

with toggle

Pipe hitch

Clove hitch

Barrel hitches: A barrel or drum is one of the most difficult objects to handle. Double check all hitches before moving such objects.

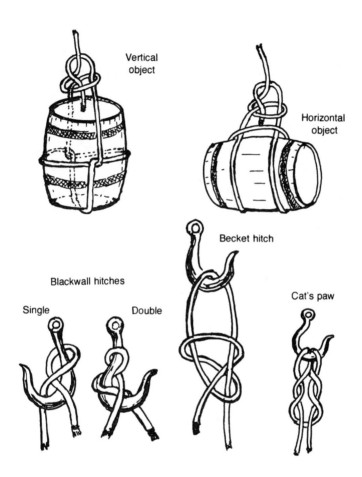

Vertical object

Horizontal object

Becket hitch

Blackwall hitches

Single   Double

Cat's paw

Scaffold hitch

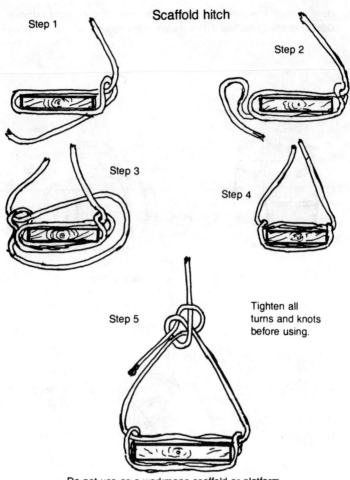

Step 1

Step 2

Step 3

Step 4

Step 5

Tighten all
turns and knots
before using.

Do not use as a workmans scaffold or platform.

Fisherman's knot

Surgeon's knot

Sheepshank

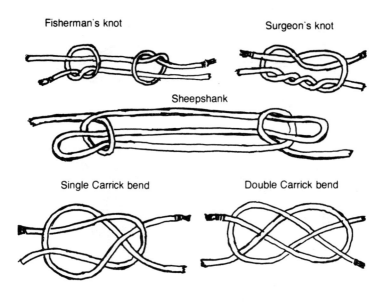

Single Carrick bend

Double Carrick bend

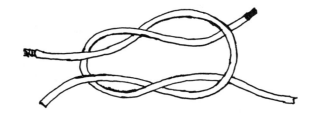

Square knot

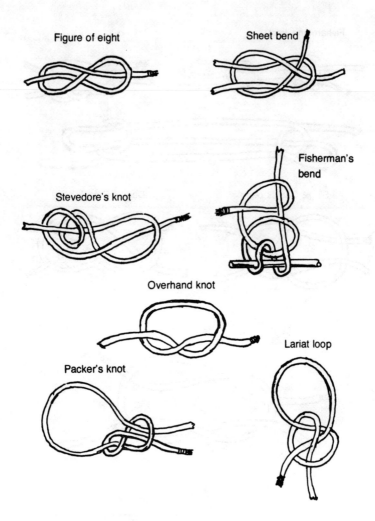

Figure of eight

Sheet bend

Stevedore's knot

Fisherman's bend

Overhand knot

Lariat loop

Packer's knot

BLOCK AND TACKLE

The use of this equipment is quite common to the plumbing industry. Many times during the coarse of construction, weights have to be lifted or moved. Following are tackle rigging combinations that will handle almost all of the moving problems.

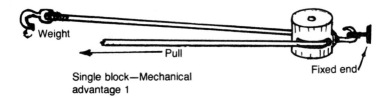

Single block—Mechanical advantage 1

Check all ropes for weight limits and for signs of wear.

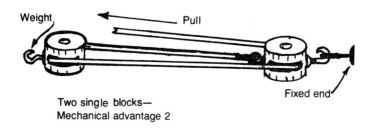

Two single blocks—
Mechanical advantage 2

Check all fixed ends for safety; lift or move only weights that rigged limits allow. Keep in the clear when moving heavy objects.

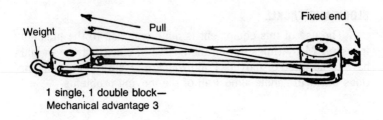

1 single, 1 double block—
Mechanical advantage 3

Where compound blocks are used, be sure that the rigging will stand any strain of mechanical advantage that combinations will permit.

Double Luff—2 fold tackle—Advantage 5

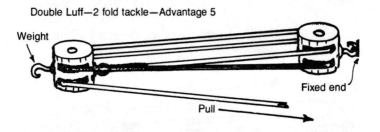

Where heavy objects are moved be sure that all workmen are in the clear. All fixed ends and all tackle should be carefully checked before moving. Never leave an object in midair.

Double Burton—Mechanical advantage 8

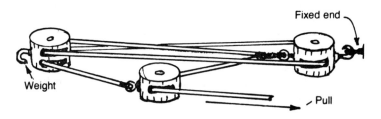

Double Burton—Mechanical advantage 11

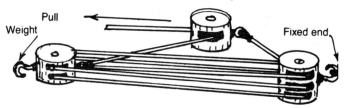

Luff on Luff—Mechanical advantage 16

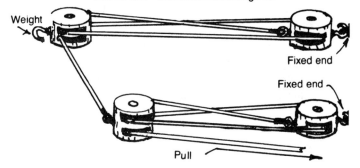

PLAN CHECKING DATA

Minimum data necessary for proper plan checking. Note: Detail all built-on-site plumbing.

1. Submit plot plan of premises and all buildings.

2. Locate all utilities furnished to site, including size and invert measurements.

3. Provide a list of all plumbing fixtures, water-using devices, and gas appliances.

4. Provide a single line piping diagram. Use proper symbols for all fixtures, devices, or gas appliances indicated on the building floor plan.

5. Show unit values of wastes, vents, and water.

6. List type of gas used and show Btu input plus gas demand in CFH for appliances to be used.

Drains and Vents

1. Show size required according to local regulation. Show changes of direction, location, and size of cleanouts, location of all drainage outlets and soil, waste and vent stacks.

2. Provide a typical elevation.

3. Indicate slope of drains and specify material.

4. Specify supports, protection, and backfill.

Water Piping

1. Size cold and hot water according to local and nationally accepted sizing practice.

2. State water pressure at site. List all pressure losses due to meters, head, and friction loss.

3. Establish extreme developed length of water pipe from meter to most remote outlet. Specify material, supports, and required protection.

4. Designate required hot water pipe size at the water heater location. Specify branch sizes.

5. Detail cross-connection control for fixtures requiring backflow devices, vacuum breakers, or air-gap fittings. Show T&P reliefs and drains.

Gas Piping

1. Show total developed length from meter to most remote outlet. Show all branches. Size all pipe.

2. Give total Btu input rating for all connected gas appliances. Give total demand in CFH.

3. Specify materials and required protection.

4. Detail piping under building slab.

Refer to accepted engineering formulas and local ordinances when submitting mechanical plan data. Example using a calculated demand of 40 units.

Drainage value .... Units	Pipe and Installation
4 toilets@ .... 6=24	Building drain........................ 4″
4 lavatories@ .... 1 = 4	Building sewer..................... 4″
2 bath tubs@ .... 2 = 4	Min. slope per ft.................. ¼″
2 showers@ .... 2 = 4	Vent as specified.
1 sink@ .... 2 = 2	Backfill as required.
1 wash tray .... 2 = 2	Materials as specified.
TOTAL —40	Test as required.

Water value .... Units	Pipe and Installation
4 toilets@ .... 3 = 12	Building supply............... ...... 1¼″
4 lavatories@ .... 1 = 4	Hot water supply .... 1″
2 bath tubs@ .... 2 = 4	Water pressure at
2 showers@ .... 2 = 4	site in psi......................... 60#
1 sink@ .... 2 = 2	Maximum distance
1 wash tray@ .... 2 = 2	from meter to most
4 hose bibbs@ .... 3 = 12	remote outlet .... 200′
Private use TOTAL—40	Materials as specified.

Note: Maximum units allowed on any branch (hot or cold) as follows: (Figured on private use)

	1″ — 25		1″ — 25	
COLD WATER	¾″ — 11	HOT WATER	¾″ — 11	
	½″ — 4		½″ — 4	

Additional loads such as sprinkler systems should be anticipated and building supply sized accordingly. Specify 1 size larger meter and supply.

Gas Appliances . . . Natural gas.	Pipe and Installation.
Furnace@ .... 110,000 Btu	Distance from meter to most remote outlet 60 feet.
Stove@ .... 82,500 Btu	
Water heater@ 33,000 Btu	Gas supply............................ 1″
TOTAL INPUT— 225,500 Btu	Material as specified
	Maximum CFH on ¾″
Demand in CFH $\frac{225,550}{1100}$ = 205	allowed................................ 105
	Maximum on ½″................. 36

DRAINAGE-UNIT VALUES

Drainage pipe sizing may be determined by assigning the following nationally accepted unit values for waste discharge. If the known discharge exceeds unit values, engineer accordingly.

Type of fixture	Minimum trap & arm size	Unit value
Bathtubs	1½"	2
Bidets	1½"	2
Dental units or Cuspidors	1½"	1
Drinking Fountains	1¼"	1
Floor Drains	2"	2
Interceptors, (grease. oil. etc.)	2"	3
Interceptors,(sand. auto wash.)	3"	6
Laundry tubs or clotheswashers, (Residential)	1½"	2
Laundry tubs or clotheswashers, (*Commercial self service laundry)	1½"*	3*
Receptors, (Floor sinks) drains from coffee urns, ice boxes etc,	1½"	1
Receptors, (Air-Gap fittings or floor sinks for commercial sinks, dishwashers, airwashers,) etc,	2"	3
Showers. (single stall)	2"	2
Showers, (gang) figure 1 unit per head. Minimum size	2"	
Sinks, (Bar, private)	1½"	1
Sinks, (Commercial) 2" waste	1½"	2
Sinks, (Industrial, washup, etc.)	1½"	3
Sinks, (Flushing rim, clinic etc)	3"	6
Sinks, Dishwashers, (Residential)	1½"	2
Sinks, (Service)	2"	3
Trailer Park traps, (1 trailer)	3"	6
Urinals, (pedestal)	3"	6
Urinals, (stall)	2"	2
Urinals, (wall) 2" min. waste	1½"	2
Urinals,(wall trough) 2" waste	1½"	3
Wash basin, (lavatories) single	1½"	1
Wash basins, (in sets)	1½"	2
Water closets, (toilets)	3"	6

*Recommend 4 units if actual GPM figures not known

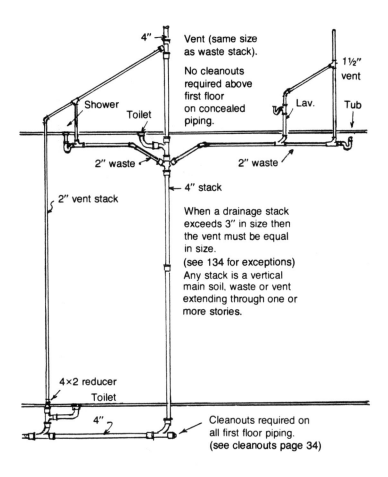

Vent (same size as waste stack).

No cleanouts required above first floor on concealed piping.

1½" vent

4"

Shower

Toilet

Lav.

Tub

2" waste

2" waste

4" stack

2" vent stack

When a drainage stack exceeds 3" in size then the vent must be equal in size.
(see 134 for exceptions)
Any stack is a vertical main soil, waste or vent extending through one or more stories.

4×2 reducer

Toilet

4"

Cleanouts required on all first floor piping.
(see cleanouts page 34)

Definition of first floor. The first floor is considered the ground floor, the slab floor on the ground, or immediately above the elevation of the ground which is intended as a floor of a structure.

TYPICAL PLUMBING INSTALLATION

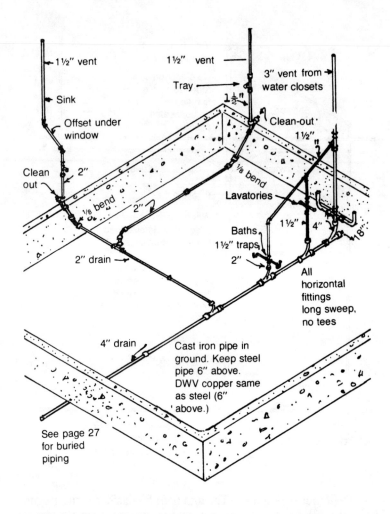

Keep 18 inches clearance in front of all main cleanouts. Have at least 12 inches clearance for waste cleanouts.

ROUGH PLUMBING COMPLETE

When calling for inspection be sure that all piping within the building is complete at the time the inspector arrives. All piping should be fully exposed; this includes water, gas, vents, drains, and any special piping that requires inspection. All valves, tub fillers, shower connections, all backing, hangers, and straps should be completed before walls are closed. Check the following points.

1. Fill the drainage system to the point of overflow.
2. Sizing of drains and vents to conform to code.
3. Heights of vent intersections.
4. Distance of vent termination from door, window, ventilators, or air intakes.
5. Distance above roof (minimum 6 inches).
6. Watertight flashings at roof, proper materials.
7. Cleanouts. Check clearance, location, size, and approved type.
8. Hangers and supports for drains for vents, straps, hangers, and supports for gas and water.
9. Pitch and alignment of waste and vents.
10. Fitting arrangement for correct drainage and venting.
11. Material standards, as specified.
12. Bath traps, shower traps, closet flanges, access doors to plumbing connections where required.
13. Steel and DWV copper drains 6 inches above ground.
14. Shower pans and subdrains in and tested as required.
15. Water pipe sized as required by code.
16. Backing, hangers, and supports in.
17. Gas pipe sized according to code. Drip tees where required. Materials as required.
18. Protection as required; no piping to be subjected to strain or stress. All pipe to be protected against corrosive action as required.
19. No piping imbedded in masonry, foundations, or walls. (Where passing through provide necessary cans and protection.)
20. Have gas gauge on, water pressure on, drains and vents full. (Have approved plans on job.)

WATER CLOSET ROUGH-INS

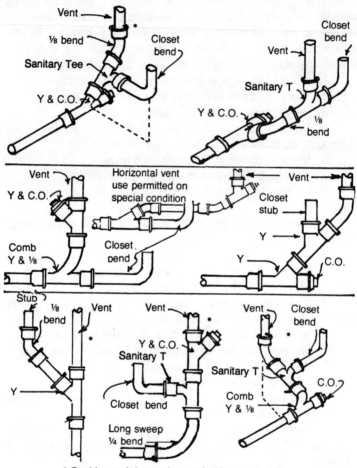

* Double rough-in may be used with approved fittings (check for allowable vent angle)

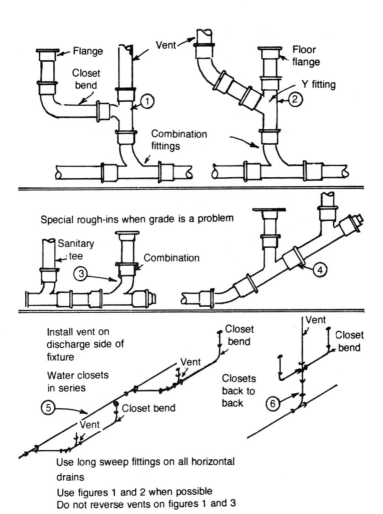

Flange

Vent

Floor flange

Closet bend

Y fitting

Combination fittings

Special rough-ins when grade is a problem

Sanitary tee

Combination

Install vent on discharge side of fixture

Closet bend

Vent

Closet bend

Water closets in series

Closets back to back

Vent

Closet bend

Vent

Closet bend

Use long sweep fittings on all horizontal drains

Use figures 1 and 2 when possible

Do not reverse vents on figures 1 and 3

165

CONSTRUCTION TRADES SCAFFOLDS

All scaffolds built to perform work that cannot be safely done on the ground must be built to safety standards as prescribed by Industrial Safety Laws. Ladders built for such work must conform to State Industrial Safety requirements. It is the duty and obligation of the employer to provide a safe scaffold of ample size and strength to perform the work intended. Following are some of the safety orders required by the Division of Industry Safety.

1. Scaffolds shall be provided for all work that cannot be safely done by workmen standing on permanent or solid construction, except where such work can be done from ladders.

2. All pole scaffolds over 75 feet high must be approved by the Division of Industrial Safety.

3. All pole scaffolds 75 feet or less in height must comply with specifications in title 8, C. A. C. Section 1603 of Construction Safety Orders.

4. All lumber used in construction of scaffolds shall be Douglas fir, not less than No. 1 common or material of equal strength.

5. Standard railings shall be provided on all open sides and ends of all built up scaffolds, runways, ramps, rolling scaffolds, elevated platforms, or other elevations of 10 feet or more above the ground or floor. Exceptions: Riveter's platforms, ladder jack, and horse scaffolds. Dangerous types of scaffolds not permitted: lean-to or jack scaffolds, shore scaffolds, boxes, barrels, loose tile, loose bricks, or other similar unstable objects shall not be used as scaffolds or to support scaffolds.

Floor openings. All holes in floors or roofs shall be covered with planks or fenced.

Ramps and runways. Ramps over 3 feet, 0 inches high for wheelbarrows shall be 2 feet, 6 feet wide, firmly cleated together and secured at each end to prevent ramp from sliding. Cleats must be provided to afford a foothold. Inclined runways minimum 20 inches wide and supported to prevent deflection. Install cleats @ 16.

STAIRS AND STAIRWELLS

1. Standard railing must be provided: a. Height—stair railing 30 inches stairwell—42 inches; b. Uprights—not less than 2 × 4 feet; c. Distance between uprights—8 feet, 0 inches.

2. Toeboards to be placed around stairwells.

3. Provide sufficient illumination.

4. Stairways and landings to be kept free from debris and loose materials, protruding nails, and splinters.

Suspended Scaffolds

1. Platforms for light trades work shall not be less than 1½ inches by 14 inches—12 foot span; 2 × 14 inches—16 foot span.

2. Provide safety line between falls. One such line for each worker.

3. Tie all hooks to a substantial object on the roof.

4. Provide back rail for all heights over 10 feet.

Standard Railings

1. Material—No. 1 Select Douglas fir or equivalent.

2. Height not less than 42 inches nor more than 45 inches.

3. Posts—2 × 4 inches spaced not more than 8 feet, 0 inches apart.

4. Toprailing—2 inches by 4 inches or equivalent, placed on that side of post that will afford greatest support and protection.

Safety Belts and Life Lines

To be provided by employer for use when it becomes necessary for workmen to crawl out on thrust-outs and such other places where no other protection is afforded them.

Horse Scaffolds

1. Lumber used in construction shall be no. 1 Select Douglas fir or equivalent.

2. Limited to 16 feet in height.

3. Distance between horses must not exceed 10 feet.

4. Platforms shall be at least 20 inches wide for light trades and 4 feet wide for heavy trades such as stone and brick masons, concrete masons, and allied trades.

5. For 10-foot spans, planks must not be smaller than 2 inches by 10 feet. Planks shall overlap 6 inches at all supports.

6. Light trades horse. Ledgers: 2 × 4 minimum for space of 4 feet or less; 2 × 6 minimum for spans between 4 feet & 8 feet.

Legs: 1 × 6 minimum for horses 4 feet or less in height; 2 × 4 minimum for horses between 4 feet and 16 feet in height.

7. Heavy Trades Horse. Ledgers: 3 × 4 minimum for spans of 4 feet or less; 2 × 6 minimum for spans between 4 feet and 8 feet.

Legs: 1 × 8 minimum for horses 4 feet or less in height: 2 × 4 minimum for horses between 4 feet and 16 feet in height.

8. Collapsible horses up to 6 feet in height may be used if strength requirements for rigid horses are observed.

Safety orders are enforced when workmen are required to operate equipment that may be considered hazardous in any way. Proper safeguards must be provided by the employer for cutting, grinding, abrasive, dust producing, chemical, explosive, paint spray, spraying devices, and or any work or equipment that will in any way cause injury or sickness to any employee.

Safety regulations are state laws provided for the benefit of the workmen and can only be enforced by cooperation of workmen and employer. Refusal to operate dangerous equipment or work in a dangerous place may lose you your job, but it might result in saving you life.

Consult safety inspectors and state laws that pertain to your occupation when in doubt.

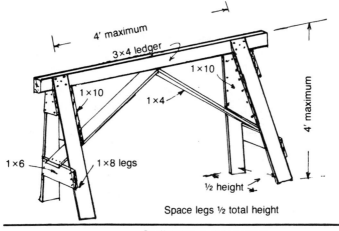

Space legs ½ total height

Scaffold horse (light trades)

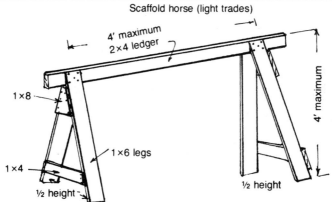

Spacing for legs must be at least half the total height of the scaffold.

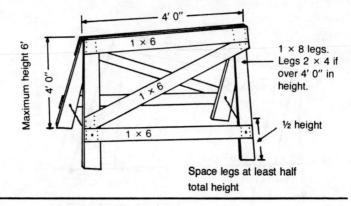

1 × 8 legs.
Legs 2 × 4 if
over 4' 0" in
height.

½ height

Space legs at least half
total height

(heavy trades)

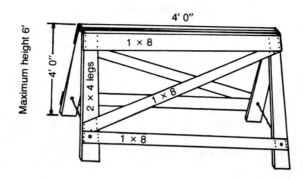

170

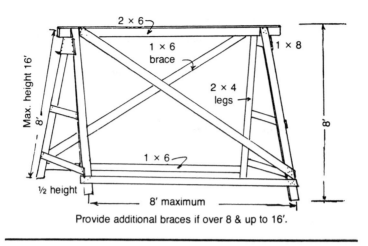

2 × 6

1 × 6 brace

1 × 8

2 × 4 legs

Max. height 16'

8'

8'

1 × 6

½ height

8' maximum

Provide additional braces if over 8 & up to 16'.

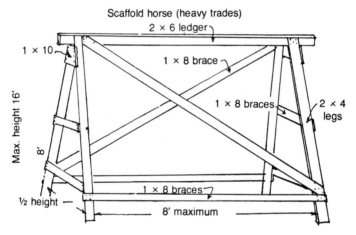

Scaffold horse (heavy trades)

2 × 6 ledger

1 × 10

1 × 8 brace

1 × 8 braces

2 × 4 legs

Max. height 16'

8'

1 × 8 braces

½ height

8' maximum

Provide additional braces for horses between 8 and 16' in height. Spacing for legs must be at least ½ the total height of scaffold.

SEPTIC TANK AND DRAIN FIELD

Inside tank size (9 feet long 3 feet wide 5 feet deep). Septic tank elbows may be used instead of baffles if desired. Minimum construction without reinforcement 5 inches thick. Locate 20 inches manholes at approximately each inlet.

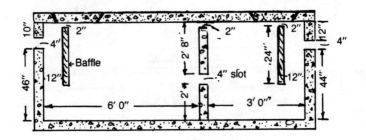

Minimum distance to house 5 feet. Solid pipe from tank to diversion box (5 feet minimum). Minimum drain field 125 feet, double for heavy soil. Minimum distance from drain line to property lines 10 feet. Keep 10 feet from any building. Locate all drains 100 feet from wells or streams. Pitch drains not more than 2 inches in 100 feet. Drain trench width 24 inches minimum. Provide 12 inches rock under pipe 1½ inches minimum size. Cover drain tile with rock 2 inches above tile. Support drains on 1 × 4 redwood board. Cover installation with building paper before backfilling.

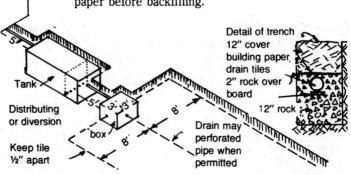

SEWER AND WATER IN SAME TRENCH

Water and sewer piping (nonmetallic) must not be installed in the same trench, except that the water may be installed if kept above the sewer at least 12 inches and placed on a solid shelf as shown below. Nonmetallic sewer pipe is any material that is not metal.

Water piping may be placed in the same trench with cast-iron or other approved metal sewer pipe if kept apart sufficiently to allow repair or to prevent galvanic or electrolitic action. It is not recommended. Check with the local inspector.

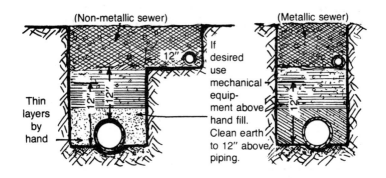

Nonmetallic sewer piping should not be installed in ground that is rocky or in soil that is improperly compacted. When such conditions occur and such piping is specified, an adequate aggregate should be used such as sand or fine gravel.

Metallic sewers should not be installed in soils that have a corrosive action on such materials. All backfill for any sewer material should be backfilled by hand to 12 inches above the pipe.

SHOWER PAN INSTALLATION

1. Liner. May be 24-gauge copper, 4 pounds of lead, or three layers of 15-pound asphalt-impregnated roofing felt. Insulate bottom of metal pans.

2. Installation. Install on sloping subfloor, minimum pitch ¼ inch per foot to approved drain. Recess into wall. Extend materials 3 inches above extreme overflow of curb, over curb, and down 1 inch. No nailing closer than 1 inch above threshold or curb.

3. Joints. Copper (silver soldered). Lead (burned) asphalt paper (fiberglass webbing 50 pound tear strength). Insulate copper and lead from other materials. Hot mopping requires complete bonding of each successive layer, including corners and turns. Caution: Do not plug weep holes in drain. Check for other approved materials.

4. Trap Size. Two inches. Vent size 1½ inches. Subdrain 2 inches with clamping ring and weep holes. Drain 2 inches.

5. Curbs and Thresholds. Minimum 2 inches and maximum 9 inches (measured from top of curb to finished drain strainer). Maximum slope on finished floor ½ inch per foot to drain.

6. Backing. Two inches thick, "solid back" behind all lining materials for substantial nailing.

7. Walls. Impervious as required by code. Install ceramic tile in showers on approved mortar bed. (Check A.S.A. 108. 1958). Walls up 6 feet minimum.

8. Doors. Safety glass or equal. Swing all doors outward.

9. Size. Minimum dimension in any cross section 30 inches. Minimum floor area 900 square inches in finish.

10. Testing. Test all showers to point of curb overflow. Check for plugged weep holes.

11. Receptors. Check for "approvals." Should be listed receptors and stalls.

12. Public Showers. Floors must drain in a manner that will prevent waste water from any bather passing over an area occupied by another.

13. Built-On-Ground Shower Receptors. May be poured monolithic as in page 177. If not install lining materials as outlined above.

SHOWER PAN INSTALLATION

Joints: Copper (silver brazed); Lead (burned).

Paper: Corners woven fiberglass strip. Materials: 24-gauge copper, 4-pound lead or three layers of 15-pound asphalt-impregnated roofing felt, hot mopped in successive layers (reinforce corners).

Test: All liners to be in at top out inspection and be leakproof under water test. Installation: All liners run up 3 inches higher than threshold and down. Provide backing for liner all around, no nails closer than 1 inch above threshold.

Minimum area for finished shower stall 900 square inches. Minimum finished interior 30 inches in any cross section. Slope finished floor not more than ½ inch to drain. Finished floor not more than 2 inches or more than 9 inches in depth (Measured to top of drain). Insulate lead or copper pans with 15-pound felt.

PUBLIC SHOWERS

All floors in public showers must be pitched so that waste water from one bather will not drain over areas occupied by other bathers. Pitch gutters ⅛ of an inch per foot.

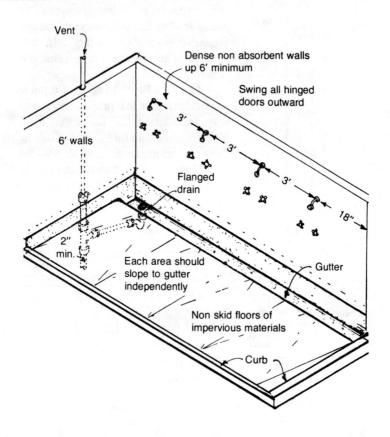

Install all gutter drains within 8 feet of side walls (16 feet maximum between drains). Round all corners to protect bathers from injury.

Minimum trap size 2 inches maximum heads allowed on 2 inches line 8. If over 8 shower heads on a 2 inches waste line, check allowable unit load.

SHOWER RECEPTORS BUILT ON
JOB SITE ON THE GROUND AS PLUMBING FIXTURES

Concrete floors and walls should be not less than 4 inches, reinforced with steel bars. Walls that have piping should be thickened to prevent weakness and cracking. Where piping is in such walls provide protection.

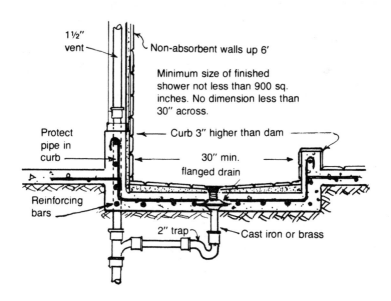

Where showers are built on the job site, the receptor should be poured monolithic with the slab. All floors and walls should be reinforced with not less than 1.2-inch steel bars on 6-inch centers, formed to extend into floor slab and up in the walls. The curb should be 3 inches higher than the threshold or dam.

SHOWER RECEPTORS

Shower receptors are a plumbing fixture and may be constructed of any material that is classified and approved for plumbing fixtures. Check local codes for approved materials. Approved receptors must be 900 square inches in area and 30 inches in any dimension.

All walls of any receptor should extend 1 inch above the threshold to ensure a leakproof joint. All walls joining the wall of the receptor must be of impervious materials and extend upward at least 6 feet.

Hinged doors must swing outward. No receptor dam or threshold should be less than 2 inches deep to the finished floor inside the receptor.

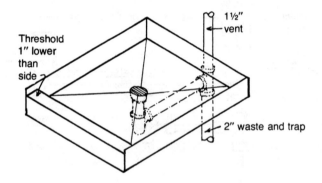

A 2-inch shower drain is cast into the floor of the receptor on some types; on others, a drain is installed similar to the installation of a sink strainer. The extension of the trap into the strainer hub is usually a caulked joint.

When installing prefabricated shower stalls, follow manufacturer's measurements and procedure.

SHOWER STALL

Typical hot mopped installation: three layers of 15-pound asphalt-impregnated roofing felt; reinforce corners with 50-pound tear-strength, fiberglass webbing.

Note: No nails in liner within 1 inch of curb.

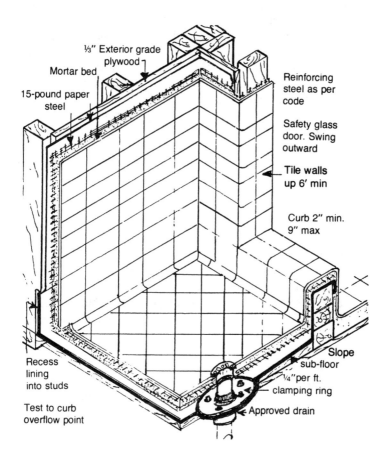

Inside finished dimension 30 inches in any cross section. Slope tile floor maximum ½ inch to drain. Slope lining material minimum ¼ inch to subdrain clamp ring.

SINK INSTALLATION (RESIDENTIAL)

1. Roughing In. Run 2 inches waste to fixture tee. Height of waste from rough floor to center of tee, 20 inches for ordinary sink with crumb cup strainer. Rough 9 inches to right or left of center of sink as desired for single compartment sink. For double compartment sink, drop waste outlet to allow for continuous waste connection. To rough in for double compartment sink 8 inches to right or left of center line of sink as desired. For garbage disposer, to sink trap rough in from floor 12 to 16 inches. Check manufacturer's measurements. Run 1½-inch vent from 2 inches waste. Where offsets are required install 45- or 60-degree fittings.

2. Water Piping. Rough in hot and cold water, 42 inches from finished floor, for wall type faucets. Hot water 4 inches to left of center line of sink, cold water 4 inches to right of center link of sink, or deck-type faucets, rough 24 inches from finished floor and 8 inches center to center same as wall faucets. Level all connections and strap securely.

3. Finish. Sink rim should be 36 inches from floor. Install strainers with putty and gaskets provided. Install faucets. Where deck type is specified, install angle valves for hot and cold water if slip joint tubing is furnished. On all connections through deck of sink make joints watertight.

4. Trap. Measure and install trap. Where tubing traps allowed, solder on approved waste bushing.

5. Water Supplies. For deck type install tubing if tubing is allowed. Do not cut tubing too short. Allow enough to remain in faucet after connection to angle valve.

6. Turn on water and check for leaks.

7. Do not use pipe wrenches on finished surfaces.

8. Grouting. Grout all connections to a finished wall.

On new work, it is best to flush out hot and cold water lines before installing faucets to get rid of all scale, grit, pipe chips or foreign matter that might damage the faucet seats.

SINK INSTALLATION

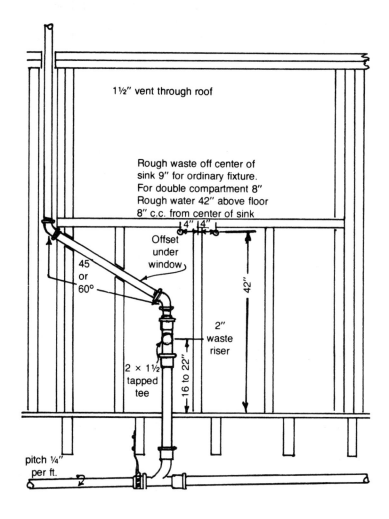

1½" vent through roof

Rough waste off center of
sink 9" for ordinary fixture.
For double compartment 8"
Rough water 42" above floor
8" c.c. from center of sink

4" 4"

Offset
under
window

45
or
60°

42"

2"
waste
riser

2 × 1½
tapped
tee

16 to 22"

pitch ¼"
per ft.

Rough waste off center of sink 9 inches for ordinary fixture. For double compartment 8 inches. Rough water 42 inches above floor 8 inches c.c. from center of sink.

Rough waste for ordinary sink 22 inches above floor. For double compartment, sink 18 inches above floor. For sink with disposer, 16 inches above floor.

SLOP OR SERVICE SINKS

Water faucets rough in wall behind slop sink back and are installed through the holes in the splash board. Always rough in to manufacturer's measurements. Strap water pipe securely to prevent movement.

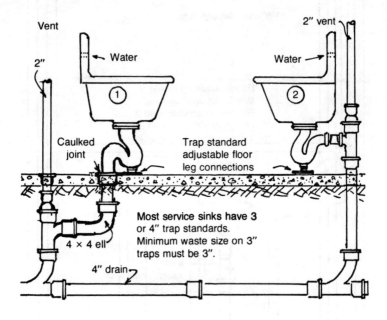

Note Where local codes prohibit 3-inch house-main drains, check with the inspection agency when 4-inch trap standards are furnished.

Some codes prohibit the "S" trap standard shown in illustration 1. Illustration 2 has the advantage of being adjustable from the floor up and from the wall out. Minimum trap size for service sinks 2 inches.

STEAM TABLE INSTALLATION

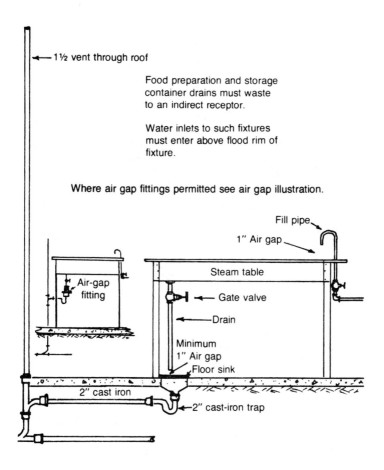

1½ vent through roof

Food preparation and storage container drains must waste to an indirect receptor.

Water inlets to such fixtures must enter above flood rim of fixture.

Where air gap fittings permitted see air gap illustration.

Fill pipe

1″ Air gap

Steam table

Air-gap fitting

Gate valve

Drain

Minimum 1″ Air gap

Floor sink

2″ cast iron

2″ cast-iron trap

Indirect receptors must be fully exposed to view at all times. Indirect receptors are prohibited unless installed in rooms of general use.

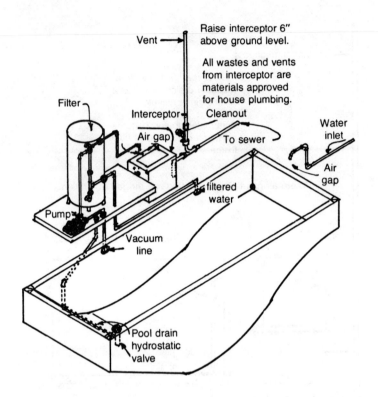

Vent → Raise interceptor 6″ above ground level.

All wastes and vents from interceptor are materials approved for house plumbing.

Filter

Interceptor → Cleanout

Air gap

To sewer

Water inlet

Air gap

Pump

filtered water

Vacuum line

Pool drain hydrostatic valve

Air breaks are required on all water supply to pool from domestic water. Air breaks are required on all wastes from backwash drains when hooked to sewer. Scum gutter drains when run separate are hooked up indirect.

Commercial or public pools piping and venting must conform to standards as outlined for structures. Private pools wastes may be used for irrigation when approved by the health department.

TESTING AGENCIES

AGA—American Gas Association
ASA—American Standards Association
ASTM—American Society for Testing Materials
AWWA—American Water Works Association
CS—Commercial Standards
FS—Federal Standards
MSS—Manufacturers Standardization Society
NBFU—National Board of Fire Underwriters
SPR—Simplified Practice Recommendations
WPOA—Western Plumbing Officials Association
UL—Underwriters Laboratories

The Western Plumbing Officials Association are the sponsors of the Uniform Plumbing Code that is widely used in western United States and part of Canada.

Gas equipment and appliances bearing the label of the American Gas Association is tested and approved for a specific use. (Check the label.)

Fire protection requirements for venting and heating appliances are checked by the Underwriters Laboratories. (Check the label or stamp.)

TRAILER PARK PLUMBING DETAILS

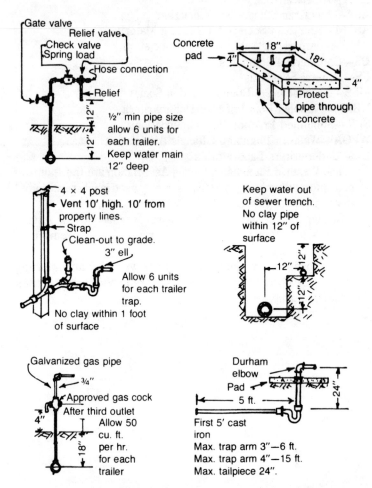

Gate valve
Relief valve
Check valve
Spring load
Hose connection
Relief

½" min pipe size
allow 6 units for
each trailer.
Keep water main
12" deep

12"
12"

Concrete
pad
18"
18"
4"
4"
Protect
pipe through
concrete

4 × 4 post
Vent 10' high. 10' from
property lines.
Strap
Clean-out to grade.
3" ell
Allow 6 units
for each trailer
trap.
No clay within 1 foot
of surface

Keep water out
of sewer trench.
No clay pipe
within 12" of
surface
12"
12"
12"

Galvanized gas pipe
¾"
Approved gas cock
After third outlet
Allow 50
cu. ft.
per hr.
for each
trailer
4"
18"

Durham
elbow
Pad
5 ft.
24"
First 5' cast
iron
Max. trap arm 3"—6 ft.
Max. trap arm 4"—15 ft.
Max. tailpiece 24".

Consult your local building department for additional rules

186

TRAILER PARK PLUMBING

See state standards for minimum grade and slope on wet vented systems and maximum allowable units.

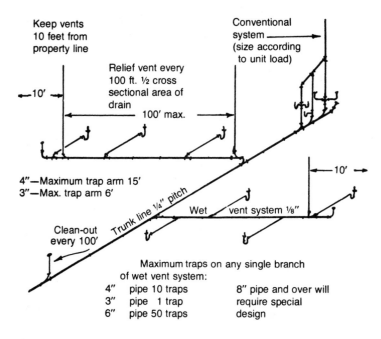

Keep vents 10 feet from property line

Conventional system (size according to unit load)

Relief vent every 100 ft. ½ cross sectional area of drain

100' max.

4"—Maximum trap arm 15'
3"—Max. trap arm 6'

Trunk line ¼" pitch

Wet vent system ⅛"

Clean-out every 100'

Maximum traps on any single branch of wet vent system:
4" pipe 10 traps
3" pipe 1 trap
6" pipe 50 traps

8" pipe and over will require special design

Maximum traps on any single branch of wet vented system:
4 inches pipe 10 traps
3 inches pipe 1 trap
6 inches pipe 50 traps
8 inches pipe and over will require special design.
Main vent for 4-inch branch, 3-inch pipe (relief 3 inches).
Main vent for 6-inch branch, 4-inch pipe (relief 4 inches).

Do not intersect any wet vented branch line with plumbing from a conventional system. Always take off house plumbing from the trunk line. ¼-inch pitch required (2 percent). Wet vented line ⅛ inch pitch required (1 percent). Trailers may be individually vented if desired. If so, size drains and vents to house plumbing standards. Check for vent and drain sizes.

TRAILER SITE PLUMBING

Locate plumbing in the rear third quarter of site. This location may extend 6 feet into the second or fourth quarter section. See cross hatch. Do not install gas piping, nonmetallic sewer drainage or vent piping, or sewage disposal units where a mobile home might be parked or a structure placed. Locate plumbing systems alongside or adjacent to lot lines. (Provide adequate support for Plumbing.)

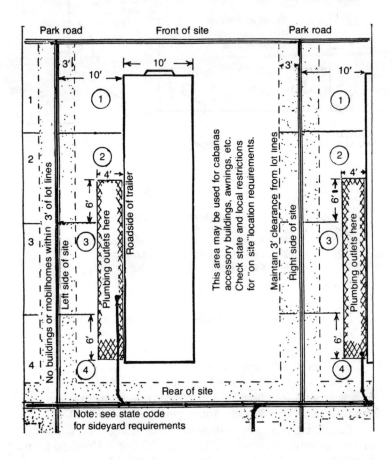

TRAILER PARK DESIGN DATA

1. Allow 6 units for water pipe sizing to each trailer site.
2. Allow 6 units for sizing drainage piping to each trailer site.
3. Allow the following demand in CFH for trailer sites in the order named (gas piping).

For the first outlet on any branch or main, (serving trailer park mobile home) 125 CFH. Second outlet on any branch or main 100 CFH. Third outlet on any branch or main 75 CFH. Allow for subsequent loadings 50 CFH.

4. Install gas piping and nonmetallic drainage and vent piping outside of areas where trailers may be parked or structures may be placed. See previous page for trailer-site plumbing.

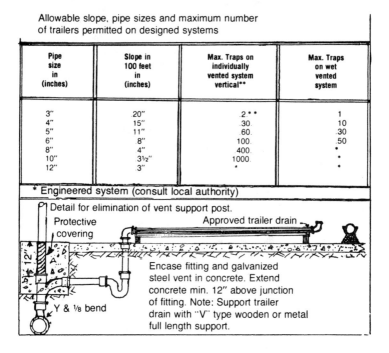

Allowable slope, pipe sizes and maximum number of trailers permitted on designed systems

Pipe size in (inches)	Slope in 100 feet in (inches)	Max. Traps on individually vented system vertical**	Max. Traps on wet vented system
3″	.20″	.2**	1
4″	15″	.30	10
5″	11″	.60	.30
6″	8″	100.	.50
8″	4″	400.	*
10″	3½″	1000.	*
12″	3″	*	*

* Engineered system (consult local authority)

Detail for elimination of vent support post.

Protective covering

Approved trailer drain

Encase fitting and galvanized steel vent in concrete. Extend concrete min. 12″ above junction of fitting. Note: Support trailer drain with "V" type wooden or metal full length support.

Y & ⅛ bend

12″

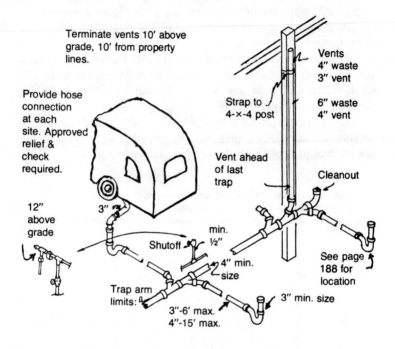

Terminate vents 10' above grade, 10' from property lines.

Provide hose connection at each site. Approved relief & check required.

Strap to 4-×-4 post

Vents
4" waste
3" vent

6" waste
4" vent

Vent ahead of last trap

Cleanout

12" above grade

3"

Shutoff

min. ½"

4" min. size

See page 188 for location

Trap arm limits:

3"-6' max.
4"-15' max.

3" min. size

Note: Mobile homes with state-approved insignia may be installed as a conventional plumbing system without traps. This system is designed as would a series of separate dwellings in a court. Venting is achieved as would the venting of the dwellings because the mobile home has approved plumbing with adequate vents. A properly designed system would take in consideration all plumbing within the mobile home as most of these have considerable more plumbing than do the small travel trailers that are not classified as independent mobile homes. Be sure to check with the agency having jurisdiction for regulations. Do not mix a system requiring traps with a system not requiring traps.

TYPES OF TRAPS

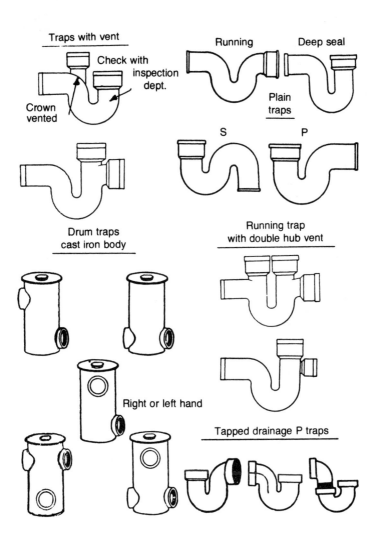

Traps with vent

Check with inspection dept.

Crown vented

Running

Deep seal

Plain traps

S

P

Drum traps
cast iron body

Running trap
with double hub vent

Right or left hand

Tapped drainage P traps

TRAP ARM LIMITS IN CONTINUOUS WASTE AND VENT SYSTEMS

All vents are measured from trap to vent along a horizontal line, except a water closet or similar fixture that is measured to include the developed distance from the top of the floor flange to the inner edge of the vent.

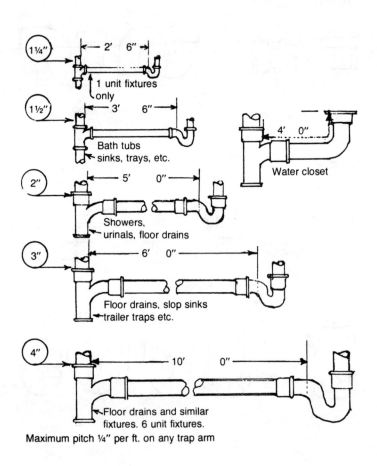

Maximum pitch ¼" per ft. on any trap arm

TRAP FEEDS

Check with local inspection agency when a trap is primed by a fixture waste.

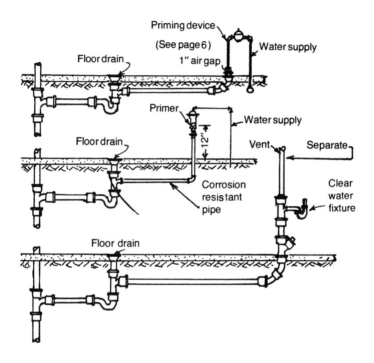

Traps that serve plumbing fixtures not in contrast use must be fed in order to maintain its trap seal. "Approved priming devices" or clear-water drainage from a plumbing fixture designated by the local plumbing inspector may be used. When installing a "fixture" feed, be sure to vent this fixture separately (see page 31). Do not attempt to improvise a "makeshift" device. Plugging an inoperative device is not the answer. Maintaining a trap seal protects health and safety in many ways.

TRENCHES

Minimum safety practices are required when trenches over 5 feet deep are dug in soils that may be considered hard and compact. Minimum timber and spacing for such soils are shown below.

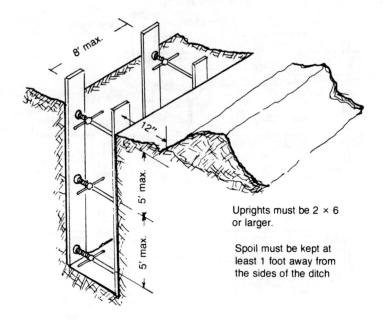

Uprights must be 2 × 6 or larger.

Spoil must be kept at least 1 foot away from the sides of the ditch

Access ladders must be provided for each 200 feet of trench. When ground conditions change, shoring and bracing to fit the condition must be provided to prevent a cave-in.

Consult the state safety orders when in doubt. Cave-ins can be avoided when all safety rules are followed.

Minimum safety practices for loose, unstable, or running materials are shown below.

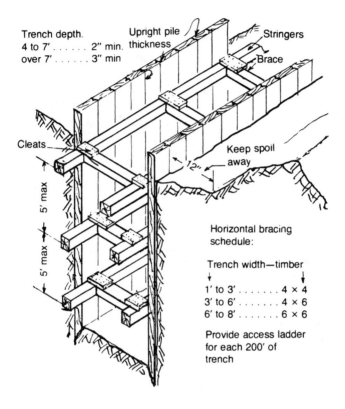

Trench depth.
4 to 7′ . . . . . . 2″ min.
over 7′ . . . . . . 3″ min

Upright pile thickness

Stringers

Brace

Cleats

5′ max

5′ max

Keep spoil away

12″

Horizontal bracing schedule:

Trench width—timber
1′ to 3′ . . . . 4 × 4
3′ to 6′ . . . . 4 × 6
6′ to 8′ . . . . 6 × 6

Provide access ladder for each 200′ of trench

Consult the safety orders when in doubt. A safe trench must be provided when workmen are employed to work in trenching below safe specified minimum depths.

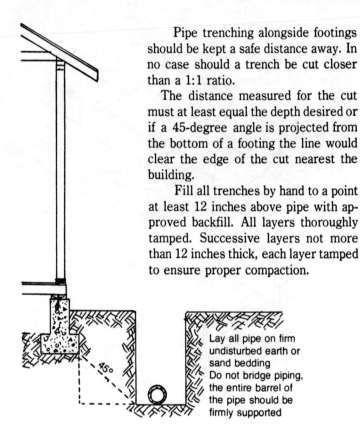

Pipe trenching alongside footings should be kept a safe distance away. In no case should a trench be cut closer than a 1:1 ratio.

The distance measured for the cut must at least equal the depth desired or if a 45-degree angle is projected from the bottom of a footing the line would clear the edge of the cut nearest the building.

Fill all trenches by hand to a point at least 12 inches above pipe with approved backfill. All layers thoroughly tamped. Successive layers not more than 12 inches thick, each layer tamped to ensure proper compaction.

45°

Lay all pipe on firm undisturbed earth or sand bedding
Do not bridge piping, the entire barrel of the pipe should be firmly supported

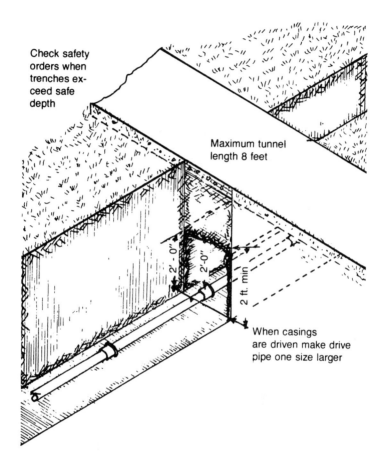

Check safety orders when trenches exceed safe depth

Maximum tunnel length 8 feet

2'-0"

2'-0"

2 ft. min

When casings are driven make drive pipe one size larger

Tunnels should be not less than 2 feet above the top of pipe (in height), and should be limited to not more than half the depth of the trench with a maximum length of 8 feet.

When casings are driven, the casing should be at least one pipe size larger than the actual outside dimension of the pipe.

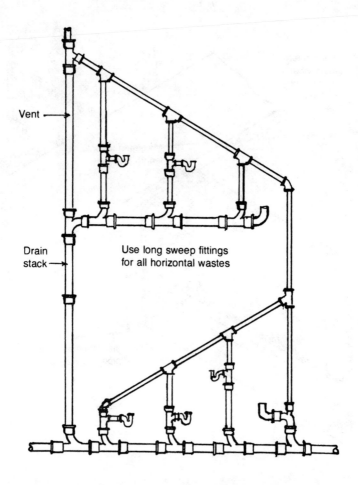

Vent

Drain stack

Use long sweep fittings for all horizontal wastes

Vent intersections must be at least 6 inches above overflow rim of fixture served by such intersection. Example: Sink 36 inches, intersection 42 inches.

TYPICAL URINAL ROUGH-INS

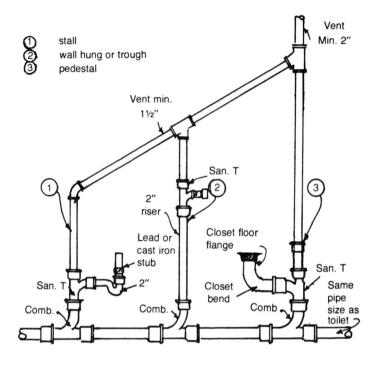

1 stall
2 wall hung or trough
3 pedestal

1. Stall urinals are roughed below the floor and are joined to the strainer of the fixture by a leaded caulked joint or by soldering a lead stub into the strainer.

2. Wall-hung urinals are roughed above the floor and may have a 1½-inch trap. Trough urinals have 2-inch traps. Check for metal carriers to support urinals.

3. Pedestal urinals rough in same as a water closet. Install urinals to manufacturer's specifications. All water supply valves for flushing must have vacuum breakers. Pipe size same as water closet.

STALL URINAL INSTALLATION

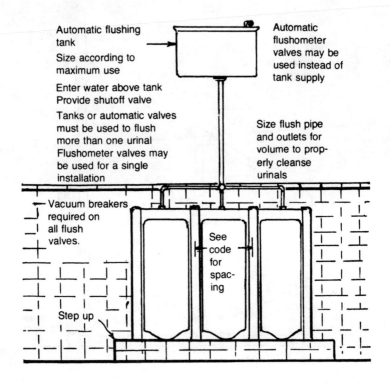

Automatic flushing tank
Size according to maximum use

Enter water above tank
Provide shutoff valve

Tanks or automatic valves must be used to flush more than one urinal
Flushometer valves may be used for a single installation

Automatic flushometer valves may be used instead of tank supply

Size flush pipe and outlets for volume to properly cleanse urinals

Vacuum breakers required on all flush valves.

See code for spacing

Step up

Two inches minimum trap size for stall urinals. One and one-half inches minimum vent size. Maximum load on single 2-inch waste with 4 stalls. Maximum load on 1½-inch vent with 4 stall urinals.

All walls and floors must be of nonabsorbent materials. Walls 4 feet high and 1 foot to each side. Extend nonabsorbent floors to 1 foot beyond the lip of the urinals.

WALL-HUNG URINALS

Check rough-in measurements for wall carriers. Walls must be at least 4 feet high and extend 1 foot to each side. Floors extend 1 foot past the urinal lip. Materials for walls and floors must be of approved, nonabsorbent construction.

Provide 30 inch spacing in stalls; 15 inches recommended to side walls. (Minimum 12 inches may be approved.)

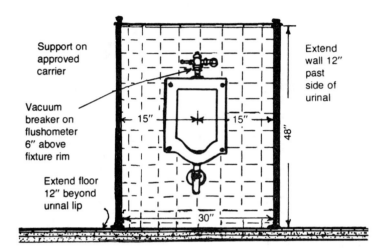

Support on approved carrier

Vacuum breaker on flushometer 6″ above fixture rim

Extend floor 12″ beyond urinal lip

Extend wall 12″ past side of urinal

15″   15″

48″

30″

Minimum riser for wall-hung urinal is 2 inches. Minimum vent is 1½ inches. Where a fixture comes into contact with the wall, a watertight joint is required. Use corrosive-resistant bolts and screws for support. When setting pull on bolts evenly. Do not strain. After testing grout to wall. Built-on-job wall or floor urinals are not permitted.

WALL-HUNG TROUGH URINALS

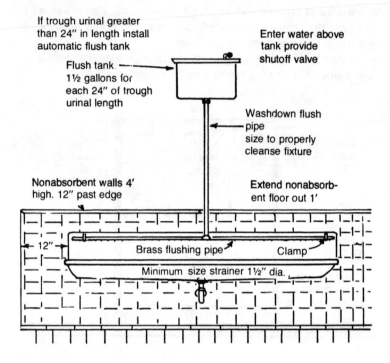

If trough urinal greater than 24″ in length install automatic flush tank

Flush tank 1½ gallons for each 24″ of trough urinal length

Enter water above tank provide shutoff valve

Washdown flush pipe size to properly cleanse fixture

Nonabsorbent walls 4′ high. 12″ past edge

Extend nonabsorbent floor out 1′

12″

Brass flushing pipe

Clamp

Minimum size strainer 1½″ dia.

Minimum waste for trough urinals 2 inches.
Minimum vent for trough urinals 1½ inches.
Minimum trap size 1½ inches.
24 inch trough equals 1 urinal, 36 inches equals 2 urinals.
48 inch trough equals 2 urinals.
60 inches equals 3 urinals.
72 inches equals 4 urinals.

Extend nonabsorbent floors out 1 foot beyond the outermost front of trough. Install a brass-perforated washdown pipe as high as possible to the back of the urinal. Allow three units for trough urinals on drainage load.

VALVES

Gate Valve. A gate valve or fullway must be used as a main shutoff valve at the building to control the water supply, at the boiler to control the hot water supply, and at all points that control more than one opening.

Compression Valves. Globe valves, angle valves, and similar shutoffs may be used to control a single outlet. All valves must be readily accessible. Each appliance that requires a hookup to the water supply must be controlled by a separate valve. Each slip-joint water connection and all nonmetallic connections require a control valve.

Valves up to 2 inches and including 2 inches must be brass or equal material. Valves over 2 inches may be cast iron or brass. All gate valves must be full way and all working parts must be of noncorrosive materials.

Check Valve. A check valve controls the directional flow of water in a pipe preventing a reverse flow due to increased pressure or other conditions.

Back-Water Valve. A type of check installed in a drain line to prevent the possibility of the sewer backing up in the house plumbing and overflowing the fixture. Drainage valves used in connection with such hookups or as shutoff valves for drainage must be full-way gate valves.

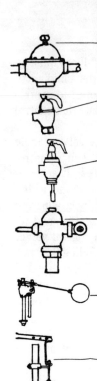

Pressure Reducing Valve. A valve used to reduce water pressure to a safe working pressure (45 to 60 psi is a normal safe water pressure.)

Pressure Relief Valve. A valve used in conjunction with a reducing valve, or in a closed system of piping to relieve excessive pressure that may build up in the piping system.

Pressure and Temperature Relief Valve. A pressure relief valve combined with a fusible plug or link that will spill at a predetermined pressure or at temperature setting.

Flushometer Valves. A valve directly connected to the water supply system and a plumbing fixture that releases a predetermined amount of water into the fixture for flushing purposes.

Float Valve. A valve used in filling tanks or vats and by the action of a lever attached to a float will close or open a water supply depending on the level of the water.

Flush Valve. A valve that controls a quantity of water from a tank and is used for flushing purposes.

Mixing Valve. A valve used in a water system to mix hot and cold water to a predetermined temperature usually in showers or public use where it may be necessary to deliver water at a safe temperature.

MAIN VENT AND SOIL STACK

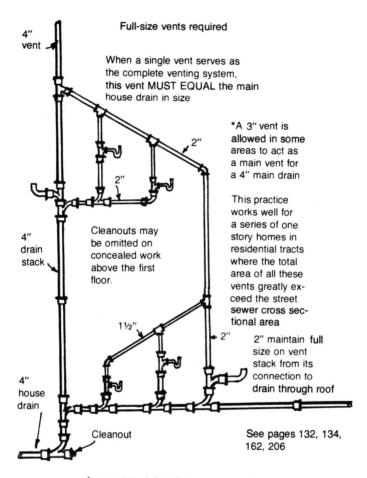

Full-size vents required

When a single vent serves as the complete venting system, this vent MUST EQUAL the main house drain in size

4" vent

*A 3" vent is allowed in some areas to act as a main vent for a 4" main drain

This practice works well for a series of one story homes in residential tracts where the total area of all these vents greatly exceed the street sewer cross sectional area

2"

2"

Cleanouts may be omitted on concealed work above the first floor.

4" drain stack

1½"

2"

2" maintain full size on vent stack from its connection to drain through roof

4" house drain

Cleanout

See pages 132, 134, 162, 206

Any vent stack installed as such shall extend undiminished in size above the roof.
*(Check with local inspection department)

VENT AND WASTE STACKS

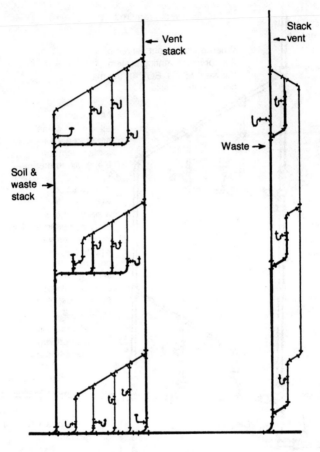

Keep vent intersections
6″ above overflow rim of
fixture served

206

VENT SYSTEMS

1 . . wet vent: waste between upper and lower T's acts as a vent pipe check with local inspection department before installing

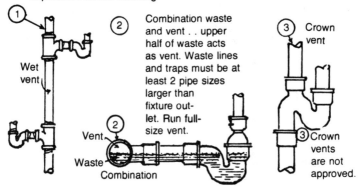

Combination waste and vent . . upper half of waste acts as vent. Waste lines and traps must be at least 2 pipe sizes larger than fixture outlet. Run full-size vent.

Wet vent

Vent

Waste

Combination

Crown vent

Crown vents are not approved.

CONTINUOUS-WASTE AND VENT INSTALLATIONS

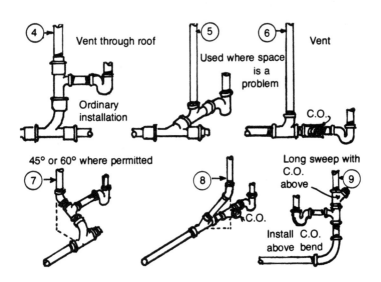

Vent through roof

Ordinary installation

Used where space is a problem

Vent

C.O.

45° or 60° where permitted

C.O.

Long sweep with C.O. above

Install C.O. above bend

FIXTURE VENT TERMINATION

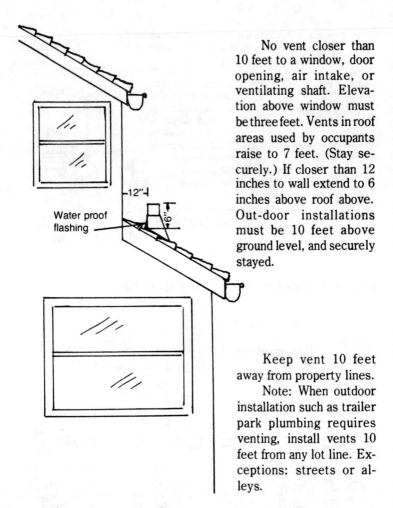

No vent closer than 10 feet to a window, door opening, air intake, or ventilating shaft. Elevation above window must be three feet. Vents in roof areas used by occupants raise to 7 feet. (Stay securely.) If closer than 12 inches to wall extend to 6 inches above roof above. Out-door installations must be 10 feet above ground level, and securely stayed.

Keep vent 10 feet away from property lines.

Note: When outdoor installation such as trailer park plumbing requires venting, install vents 10 feet from any lot line. Exceptions: streets or alleys.

YOKE VENTING

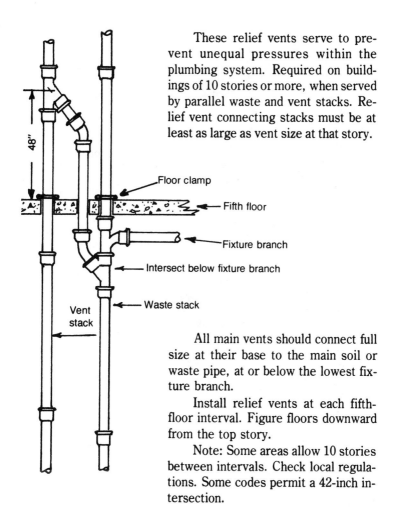

48"

Floor clamp

Fifth floor

Fixture branch

Intersect below fixture branch

Waste stack

Vent stack

These relief vents serve to prevent unequal pressures within the plumbing system. Required on buildings of 10 stories or more, when served by parallel waste and vent stacks. Relief vent connecting stacks must be at least as large as vent size at that story.

All main vents should connect full size at their base to the main soil or waste pipe, at or below the lowest fixture branch.

Install relief vents at each fifth-floor interval. Figure floors downward from the top story.

Note: Some areas allow 10 stories between intervals. Check local regulations. Some codes permit a 42-inch intersection.

VERTICAL WASTE INTERSECTIONS

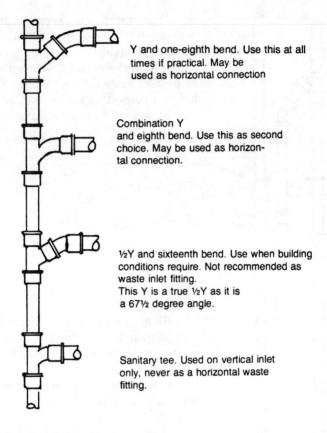

Y and one-eighth bend. Use this at all times if practical. May be used as horizontal connection

Combination Y and eighth bend. Use this as second choice. May be used as horizontal connection.

½Y and sixteenth bend. Use when building conditions require. Not recommended as waste inlet fitting.
This Y is a true ½Y as it is a 67½ degree angle.

Sanitary tee. Used on vertical inlet only, never as a horizontal waste fitting.

Properly designed waste fittings will increase flow of waste, promote scouring action, help prevent stoppage, and will ensure a better installation.

MINIMUM SPACE FOR WATER CLOSETS

Water closets and urinals must be allowed sufficient space to be properly installed and maintained. Allow 30 inches for minimum space center to center. Do not rough closer than 15-inch finish to the nearest side wall.

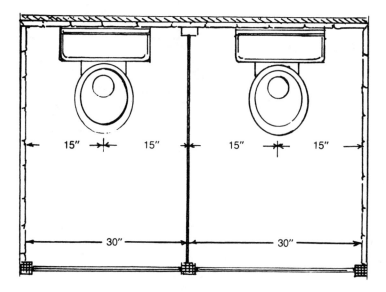

All water closets must have at least 15 inches clearance from the space center to the nearest side wall. All compartments must be at least 30 inches wide in the finished wall dimension. Clearances in front depend on door arrangement and is usually around 60 inches.

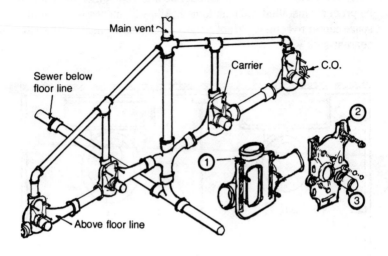

Wall-hung water closets are roughed in above the floor line. Check manufacturer's measurements. Number 1 in the detail above is the body of the fitting installed in the line. Normal grade is maintained as the opening in the body of the fitting allows for normal grade. On long runs, check the fitting allowance for the necessary required pitch.

Number 2 is the carrier that the toilet is bolted to and, in turn, is bolted to the body. Note that the carrier is tapped for a 4-inch thread. In bolting the fittings together, install the gasket provided for the opening.

Number 3 is the adjustable nipple provided for the toilet bowl connection. Each nipple has a locknut and a running thread that will adjust to a desired length. When installing make all bolts up evenly. Do not strain.

WALL-HUNG WATER CLOSETS (TANK TYPE)

Support as specified by manufacturer.

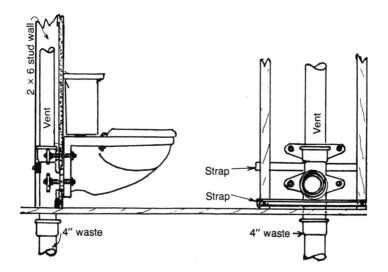

Wall and floor carriers for fixture vary with the manufacturer. A wall-hung toilet developed in the past few years allows the bowl to be elevated above the floor line. This bowl connects to a fitting that is secured to the stud wall by straps front and rear. It roughs in an ordinary 2-×-6 stud wall.

The fitting and carrier is screwed or bolted to the stud and the bowl, in turn, is bolted to the carrier. An adjustable nipple and locknut is screwed into the fitting to receive the bowl. In mounting the bowl to the carrier, care must be used; it is possible to break the bowl if an uneven pressure is applied. Always check manufacturer's specifications before roughing in because this required accurate measurements.

WASH BASIN INSTALLATION

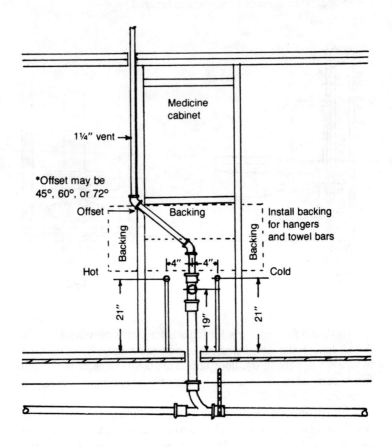

For an ordinary wash basin, rough in waste at 19 inches. For a basin with pop-up waste, rough at 16 inches. Rough water 2 inches above waste and 4 to 6 inches off center of waste. For special basins check rough-in sheet. Install backing for hangers and towel bars.

* 72 degrees allowed in some localities.

WASH BASIN OR LAVATORY INSTALLATION

    1. Support. Install ample backing to properly support basin on wall. Run backing to receive towel bars and legs. For wood backing, run 2-inch thick lumber. Metal supports fasten as directed by the manufacturer.

    2. Roughing In. Install 1½ inch minimum waste. Install 1¼ inch minimum vent. Hot and cold water pipe ½ inch minimum size. Height of waste to center of Tee to rough floor 20 inches for ordinary P.O. plug. For pop-up waste 16 inches. Water pipe rough, allow 2 inches above waste outlet. Install hot water 4 inches to left of waste, cold 4 inches to right of waste outlet. For special fixtures, follow rough-in measurements furnished by manufacturer.

    3. Finish. Install hanger to finish rim of fixture 31 inches from finished floor. Install wall flanges, cover tubing, and angle valves.

    4. Basin Trim. Install faucets, where faucet connects to basin make leakproof connection with gasket or putty. Faucets must be approved type with spout outlet 1 inch above fixture rim. Install P.O. plug or pop-up waste assembly. Make water tight connection with putty and gasket.

    5. Hanging Basin. Hang to hanger provided and level.

    6. Trap. Install 1½ inches approved trap. Measure to center of tailpiece, solder on approved bushing if tubing trap allowed, screw on proper nipple if screw trap furnished, install wall flange or cover tubing, and join to tailpiece.

    7. Water Supplies. Cut tubing to join angle stops. Where slip joints are allowed, cut tubing long enough to slip into angle valve and still remain into the faucet.

    8. Grouting. Complete job by filling all joints to wall with approved grouting. Turn water on and check for leak.

    9. Do not use pipe wrenches on any finished surfaces.

WASH TRAY INSTALLATION

1. Roughing In. Install minimum horizontal 2 inch waste line. Waste riser may be 1½ inches if desired. If 1½-inch riser is used, enter 2 inches waste with proper reducing fittings. Rough waste 12 to 16 inches. Install 1½ inch vent. Rough in water 40 inches above finished floor line 8 inches center to center, hot 4 inches to left of center of tray, cold 4 inches to right. Level all connections and strap securely.

2. Supports. Install backing for tray hangers or brackets.

3. Materials. Check with local inspection department for tray materials. Cement trays may be prohibited.

4. Finish. Fasten hanger or brackets to wall to receive tray. Install faucets minimum clearance of spout of faucet 1 inch above rim of fixture. Where deck-type faucets are installed, make a leakproof joint. Level fixture to wall. Install strainer and tail-piece connection. Use putty and gasket provided where strainer joined to tray.

5. Turn water on and check for leaks.

6. Grout where joined to wall.

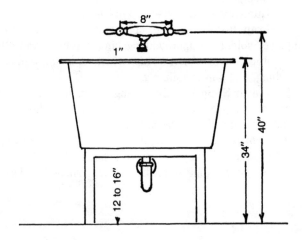

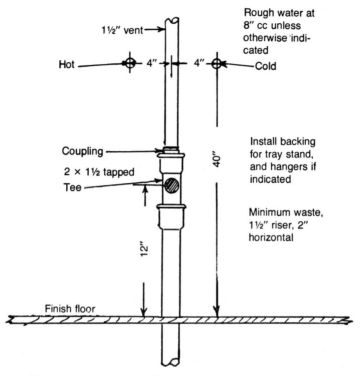

Rough water at
8″ cc unless
otherwise indi-
cated

1½″ vent

Hot — ⊕ — 4″ — 4″ — ⊕ — Cold

Coupling

2 × 1½ tapped

Tee

40″

12″

Install backing
for tray stand,
and hangers if
indicated

Minimum waste,
1½″ riser, 2″
horizontal

Finish floor

Measurements for tray finish at 34 inches. If a different level is desired, raise or lower by adding or subtracting the difference desired.

WATER PIPE INSTALLATION

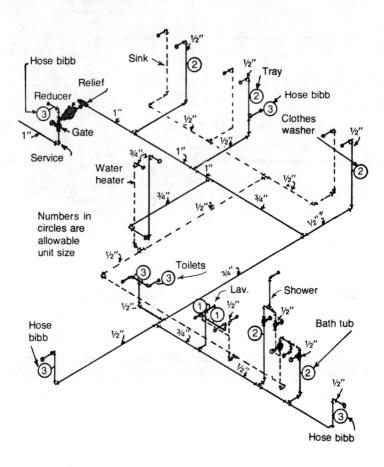

Typical two-bath job. Length from meter to most remote outlet 60 feet. Pressure 46 to 60 pounds. If pressure is over 100 pounds, install reducing valve and relief as shown. If a sprinkler system is installed, increase service line. Add for each head 1 unit or figure demand in gallons per minute. Install gate valves at main house supply and at inlet to water heater. Recommend ¾ inch to hose bibbs.

MAXIMUM LOADS IN UNITS FOR WATER PIPE

Meter & building →		(L) = 30 to 45 psi. S = Supply (M) = over 45 to 60 (recommended range) (h) = over 60 to 80 (recommend reducer)							
		Maximum Units according to pressure in column below distance in feet							
		50′	75′	100′	150′	200′	300′	400′	500′
¾″ meter S = ½″	(L) (M) (H)	6 9 11	5 8 9	4 6 7	3 5 6	2 4 5	0 0 0	0 0 0	0 0 0
¾″ meter S = ¾″	(L) (M) (H)	.18 27 34	16 23 28	12 17 22	9 14 17	6 11 13	0 8 10	0 6 8	0 5 7
1″ meter S = 1″	(L) (M) (H)	36 60 87	31 47 66	25 36 48	20 30 38	17 25 32	13 20 26	12 18 22	10 15 19
1½″ meter S = 1¼″	(L) (M) (H)	90 168 237	68 130 183	48 89 127	38 66 93	32 52 74	25 39 54	21 33 43	19 29 37
1½″ meter S = 1½″	(L) (M) (H)	151 270 366	124 225 311	91 167 240	70 128 186	57 105 154	45 68 113	36 62 88	31 52 73
2″ S = 2″	(L) (M) (H)	372 570 690	329 510 670	265 430 560	217 368 478	185 318 420	147 250 340	124 205 288	107 173 245
2″ meter S = 2½″	(L) (M) (H)	445 680 690	418 640 690	370 580 690	330 535 690	300 500 650	265 440 570	240 400 510	220 365 460

To read the chart, first determine the distance from the source to the most remote outlet. Total units are allowed for a given pressure as shown below the distance. For distance over 200 feet and units in excess of 50, it may be wise to design as outlined in BMS 66 (United States Government Printing Office). Units are figured to combine hot and cold water demand.

ALLOWABLE UNITS ON BRANCH LINES

Branch Line Size ↓	All branches off the length of supply are assigned a unit value. (B is for branch)							
	Developed distance supply to remote outlet							
	50'	75'	100'	150'	200'	300'	400'	500'
B=½" (L)	6	5	4	3	2	0	0	0
(M)	9	8	6	5	4	0	0	0
(H)	11	9	7	6	5	0	0	0
B=¾" (L)	18	16	12	9	6	0	0	0
(M)	27	23	17	14	11	8	6	5
(H)	34	28	22	17	13	10	8	7
B=1" (L)	36	31	25	20	17	13	12	10
(M)	60	47	36	30	25	20	18	15
(H)	87	66	48	38	32	26	22	19
B=1¼" (L)	90	68	48	38	32	25	21	19
(M)	168	130	89	66	52	39	33	29
(H)	237	183	127	93	74	54	43	37
B=1½" (L)	151	124	91	70	57	45	36	31
(M)	270	225	167	128	105	68	62	52
(H)	366	311	240	186	154	113	88	73
B=2" (L)	372	329	265	217	185	147	124	107
(M)	570	510	430	368	318	250	205	173
(H)	690	670	560	478	420	340	288	245

Use same unit value for hot or cold branch sizing. In sizing either one, use the column that gives the distance from source to most remote outlet of the cold water line. Note: Use same distance as determined for length of supply. When using tables it should be understood that factors are considered to fit all conditions. When pressure-reducing valves are used, figure from the location of the pressure-reducing valve to the most remote outlet. When using tables be sure to use the known pressure.

(L) = 30 to 45 psi
(M) = over 45 to 60 psi (recommended range)
(H) = over 60 to 80 psi (reduce as code requires)

WATER PIPE SIZING (RESIDENTIAL USE)

Distance from meter to rear bibb 200 (M) pressure. Fixture unit allowance combines hot and cold demand. Adding combined total of all sections equals 50 units.

See meter and building supply. Find under 200-foot column. Fifty units will require a 1½-inch meter and a 1¼-inch building supply. Unit allowance includes hot and cold demand. Starting from rear bibb and working toward the meter add units until total exceeds limit that is allowed for size limitations.

Using the branch line size in the 200-foot column, work toward source; increase as unit allowance exceeds branch limits.

Allowing 4 units on ½-inch pipe, 11 on ¾ inch pipe, 25 on 1-inch pipe, and 52 on 1½-inch pipe. Size sections A(9) units until junction with sections (B) 4 units. Because 13 units exceed limits on ¾-inch pipe, increase at that total again is 22, which allows 1-inch pipe. At junction of sections D, total for all sections is 50 requiring 1¼ inches supply.

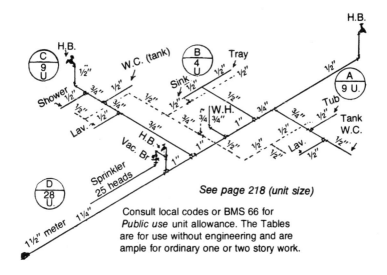

See page 218 (unit size)

Consult local codes or BMS 66 for
Public use unit allowance. The Tables
are for use without engineering and are
ample for ordinary one or two story work.

SIZING WATER PIPING FROM TABLES

The following rules must be followed to properly size water piping when tables are used. It must be kept in mind that the tables are limited in scope. In tables, all variables are allowed for up to the limits of pressure, distance and allowable unit demand. (See sizing from graphs.)

1. Determine distance from meter or other determined source to most remote outlet to be supplied. (This is the only distance used.)

2. Tabulate unit list according to private or public use. (See page 50 for definitions.)

3. Determine pressure available at source. Allow for elevation. Example: If the elevation is 45 feet above the meter, subtract (45 × .43) or 19.4 psi from pressure at meter. Additoinal losses such as meter or other devices that may account for friction or pressure loss must be considered.

4. Select a pressure range that corresponds to the final determined available pressure.

5. Select column corresponding to determined distance. Use the same distance for supply or branch.

6. Knowing the total units, find the corresponding units allowed in the pressure range selected. Size of building supply will be shown in supply column.

7. To find branch size, start at the most remote outlet and, using the same distance as established for building supply, work back toward the source of the supply. When units exceed allowable pipe size, increase to next size at this junction. (See page 221.) The main building supply and branches are now determined.

8. Hot water sizing from tables. Use same unit value and same distance as determined for building supply. Start at the most remote outlet of hot water, work toward the water heater until units exceed allowable pipe size. Note: No decrease in fixture unit value for hot water is allowed when sizing from tables. Building supply size at the junction of the water heater is computed by adding the additional load imposed by the hot water unit demand at this point.

WATER PIPE SIZING BY THE UNIT SYSTEM

A unit as used in sizing waterpipe is an arbitrary figure that relates to a theoretical demand. Charts and graphs have been developed that will give a fairly accurate measurement of water use. When used correctly, the unit system will give adequate calculations to properly size water pipe. Note: See private and public use unit allowances.

Fixture	Unit Value (Private)	Unit Value (Public)
Bar Sink	1	2
Bathtub (includes shower over)	2	4
Dental Unit or Cuspidor		1
Drinking Fountain (each head.)		1
Hose Bibb or Sill Cock	3	5
House Trailer (each)	6	6
Laundry Tub or Clotheswasher (Each pair of faucets)	2	4
Lavatory	1	2
Lavatory (Dental)	1	1
Lawn Sprinklers (each head)	1	1
Shower (each head)	2	4
Sink (Service)		4
Sink or Dishwasher	2	4
Sink (flushing rim or clinic)		10
Sink (washup, each set of faucets)		2
Sink (washup, circular spray)		4
Urinal (pedestal or similar type)		10
Urinal (stall)		5
Urinal (wall).		5
Urinal (flush tank)		3
Water Closet (flush tank)	3	5
Water Closet (flushometer)	6	10
Note: When using Tables to size systems serving water closet *flushometers*, size from most remote value. Be sure to assign the following values.		
First flushometer		40
Second flushometer		30
Third flushometer		20
Fourth flushometer		15
Fifth and all others		10

SIZING WATER PIPING FROM GRAPHS

After constructing the tabulation form shown below, eight steps must be considered. The example will consider a two-story office building. The length of building supply 200 feet. Highest use 25 feet; water pressure at meter 55 psi.

Fixture Use (public) Fixture List	Building Supply		Hot Water Supply	
	Total Number of Fixtures	Total Unit Value	Number of Fixtures Using Hot Water Only	Unit Value at 75% or ¾ Total Unit Value
Water Closet (flush tanks)				
Water Closets	130	650		
Urinals	47	235		
Service Sinks	27	108	27 × 4 × ¾	81
Shower Heads	12	48	12 × 4 × ¾	36
Lavatories	130	260	130 , 2 × ¾	195
Total Units		1.301		312
Gpm Demand (see dotted line) page 226		248		88
Pipe Size (page 227)		3″		2″
See next page for step by step procedure to work the example given on this page. 8 steps outlined				

See next page for step-by-step procedure to work the example given on this page.

1. Construct tabulation chart. (Find total units.)

2. Find demand in GPM (See chart A, page 226.)

3. Check site pressure, subtract losses, multiply by 100 (Loss per 100 feet). Example on page

4. Determine greatest length to most remote outlet using water. Allow for fitting.

5. Divide result from line three by result of line four. (This will give ordinate line in graph.)

6. Determine type of pipe to be used. (If local conditions are such that liming or caking may occur, select graph to correspond.)

7. Select graph as determined by local condition.

8. At bottom of graph, select friction loss line as determined by line five. From sides of graph select a line that corresponds to known GPM. Where lines intersect, find required size.

SIZING WATER PIPE FROM GRAPHS (EXAMPLE)

Use tabulation data from preceding page (data to be considered as follows).

Steps 1 through 8. Length 200 feet. Pressure 55 psi. Height 25 feet.

1. Maximum length including fitting allowances 200 feet. (Measurement includes to highest point.)

2. Pressure available at building site, 55 psi.

3. Total units. 1301 (from tabulation chart). All fixtures demand in GPM. See dotted line on 226, 248 GPM. Hot water same chart 88 GPM.

4. To determine pressure available for friction loss, consider the following factors.

a. Total delivered pressure at site, 55 psi.

b. Allow for head loss due to elevation of upper fixtures 25 feet. (25 × 0.43) = 10.75 psi.

c. Allow 8 psi for flushtank system. This demand is a reserve to prevent pressure drop below this point when maximum use occurs.

d. Allow meter loss for flow through 3-inch meter of 248 GPM. (See dotted lines page 226) = 17 psi.

e. Subtract total losses 10.75 + 8 + 17 from 55 leaves a working pressure of 19.25 (psi).

f. To find allowable friction loss per 100 feet, multiply working pressure in this case 19.25 by 100 = 1925. Divide by length (in this case 200 feet). Result is 9.62 or 9.6. This will be the ordinate line used in this example,

5. Select graph to suit local water characteristic. In this example use fairly smooth. See dotted lines, page 226.

6. Follow dotted line upward to point of intersection of cross line of GPM, demand find 3-inch pipe. Note: If lines intersect between pipe sizes always go to the next larger size.

7. To find water branch size, use same ordinate line follow to intersecting GPM crossline. Find 2-inch pipe for this example.

8. Size branches as outlined. For uses not covered by the unit, allowance method always use the actual demand in gallons per minute.

Note: Due to the many variables encountered in water demands, valves to control these factors may be installed and adjusted to equalize sections.

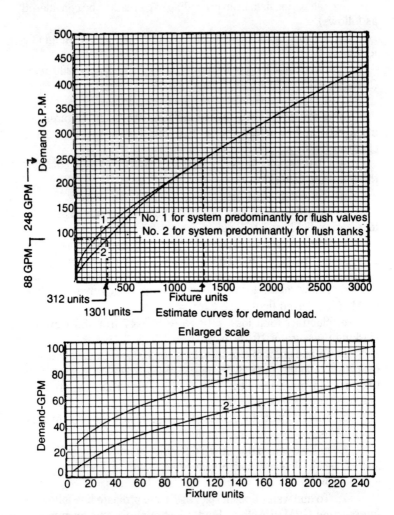

Estimate curves for demand load.

Enlarged scale

Note: When designing make sure to allow for flush valves. Notice the difference on the enlarged scale demand. Allow fixture units as directed by local code requirements.

SIZING WATER PIPE FROM GRAPHS

The fairly smooth graph will be used for new steel pipe with no anticipated interior buildup. Friction loss in head in psi for 100 feet length.

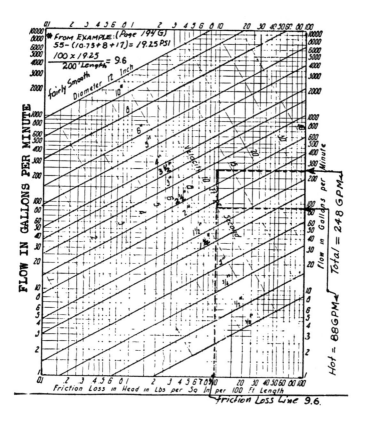

Note: Vertical lines are friction-loss lines. Always multiply "pressure available" for friction loss by 100. Divide this figure by the developed length to find friction loss line. Example: if length were 385 feet, loss line would be five.

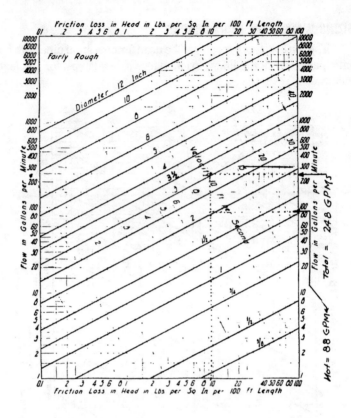

Check dotted lines with rough graph. Note: Use this graph when anticipating some liming, caking, or other interior buildup. Loss of GPM should always be considered when designing any water system. The fairly rough graph will adapt to certain known conditions.

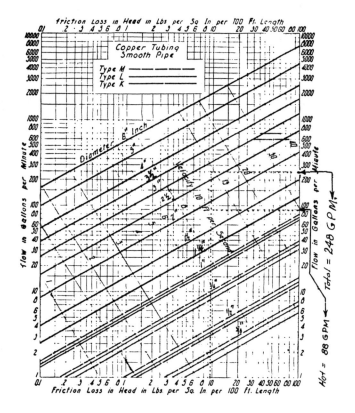

Check code regulations for installation. Note: Notice the three lines in the smaller-size pipe. This is due to the interior loss in size due to the wall thickness. Use the smooth graph when no interior buildup is anticipated.

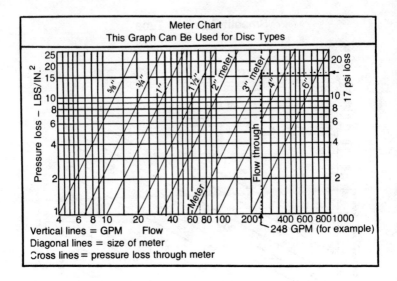

Meter Chart
This Graph Can Be Used for Disc Types

Vertical lines = GPM   Flow
Diagonal lines = size of meter
Cross lines = pressure loss through meter

248 GPM (for example)

Check the following points before sizing pipe.

1. Type of pipe and material to be considered.

2. Pressure to be delivered at job site.

3. Whether job is to be metered or a direct connection provided.

4. Characteristic of water in the locality.

5. Allowances for meters, valves, fittings, elevation, length of run, maintaining of a minimum pressure when on maximum demand and for water softeners and other devices that may account for head or friction loss.

6. When sizing the building supply, it is good practice to increase the service pipe one size.

7. Note meter size on graph above. If losses are crucial, it would be wise to go to the next larger size meter and service.

8. Consult the local utility before designing in localities are unfamiliar.

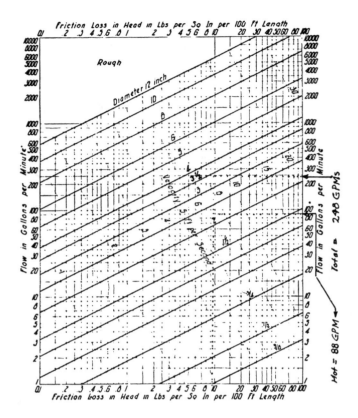

Friction Loss in Head in Lbs per Sq In per 100 Ft Length

When local conditions are such that interior buildup is expected, this chart will provide for this by the increase in size. Compare this to the fairly smooth graph. Notice this will take a 3½-inch pipe with no margin of supply. This pipe should be 4 inches if extreme buildup occurs.

WATER HAMMER

Water hammer is a condition that occurs in a water supply pipe to fixtures or appliances controlled by valves. A quick closing valve upon suddenly shutting off will create a condition of shock resulting in an audible sound called water hammer.

To help control this condition several things can be done. An effective way is to install a larger pipe than the supply at a predetermined point, installed in a vertical position and as far above the supply valves that conditions permit. This pipe is called an air chamber and is normally around 24 inches long with the end capped. The pipe filling with water compresses the air, resulting in a so-called air cushion. This installation is satisfactory only as long as the air is retained in the chamber. Repeated use gradually forces all the air out and is replaced with water. Provisions must be made to replace the air. Install all air chambers in a manner that will allow them to be drained at intervals.

Several types of shock absorbers are on the market that are designed to absorb the shock. These devices are installed as part of the water system and are guaranteed by the manufacturer to do certain things as specified. Consult the manufacturer's specifications for size and length of piping installation.

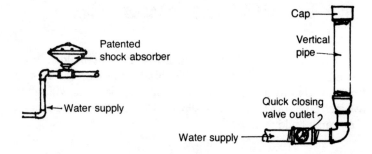

WATER HEATER INSTALLATION

1. Locations. Do not install water heaters in bedrooms, bathrooms, or closets of such rooms. Do not locate under interior stairways or areas where there may be storage of combustible or explosive materials.

2. Flues. Independent flues recommended for each appliance. Where allowed in manifold see page 237. Check department regulations on "code" when combining with other type gas appliances.

3. Materials for Flues. Single wall (class C) permitted inside only. Must be exposed entirely to view. Provide 6 inches clearance to materials considered combustible. Double wall or cement asbestos (class B). Provide 1-inch clearance to combustibles. (Check for added clearances.)

4. Connections. Enter on "T" inlet for exterior flues. Support interior flues with proper hangers or brackets. Do not support on appliance.

5. Gravity Flue Termination. Minimum 12 inches above roof. Keep 4 feet away from windows, doors, air intakes, ventilating shafts, or openings. Extend 12 inches above such openings and provide proper support.

6. Vent Caps. Check for approved caps. Do not allow flue to project into cap venting area.

7. Water Connections. Install gate valve or fullway at cold water inlet. Provide unions at both hot and cold connections. Install reliefs as shown. (See page 234.)

8. Gas Connections. Install approved gas cock or approved appliance connector. All shutoffs to be readily accessible. Rigid connections required on appliances 100,000 Btu or over. (Check for A.G.A. approvals on gas connectors.)

9. Air and Ventilation. Minimum air to any water heater compartment 100 square inches plus 1 square inch for each 1000 Btu. Provide air in equal parts tops and bottom. Minimum ¼-inch wire mesh for screens.

10. Approvals. Do not install any water heater that does not have a label of approval from a nationally recognized testing agency.

11. Notes. Keep 10 feet away and 3 feet above any forced air intake with gas-appliance vent termination.

Water Heater Installation

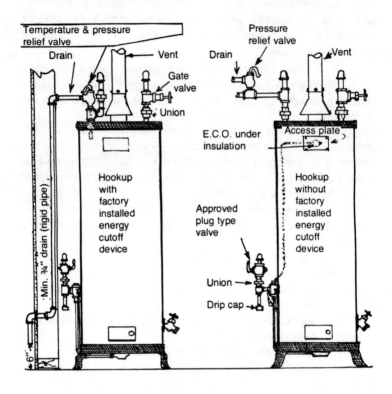

1. Use approved relief valves. Check for A.S.A. rating and capacity. Run full-size drain to outside within 24 inches and 6 inches off the ground. Do not thread end of drain. Do not trap drain.

2. Provide 100 square inches combustion air plus 1 square inch for each 1000 Btu to any water heater in closets or confined areas.

3. Elevate water heaters in garages a minimum of 24 inches above the floor. Provide permanent protection.

4. Install T & P valves in the hottest point of tank. Use extended type only.

5. Install pressure valves as directed by code.

WATER HEATER INSTALLATION IN ATTIC

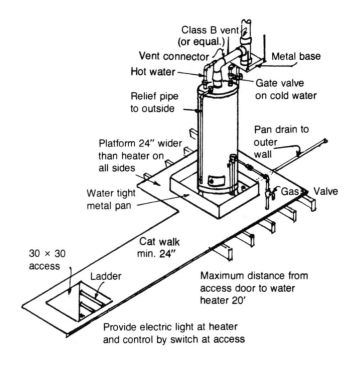

Class B vent (or equal.)

Vent connector

Hot water

Relief pipe to outside

Metal base

Gate valve on cold water

Platform 24" wider than heater on all sides

Water tight metal pan

Pan drain to outer wall

Gas Valve

30 × 30 access

Cat walk min. 24"

Ladder

Maximum distance from access door to water heater 20'

Provide electric light at heater and control by switch at access

Maximum distance from access door to water heater 20 feet. Provide electric light at heater and control by switch at access. All water heaters installed in attics should be installed on adequate sized joist to fully carry the weight required.

Where it is necessary to install a water heater in an attic, a metal watertight pan and drain should be provided to take care of a possible leak that might develop.

All vents in attic must be class B vents and clear combustible materials as code requires.

1. Base. Solid floor 24 inches larger than heater on all sides.

2. Access. Permanent ladder or stairs to access door large enough to remove heater if required. Provide 24 inches cat walk to heater location.

3. Flue. Extend class B flue through roof at least 12 inches above and 4 feet away from any wall, window, door, ventilator, or air intake. Install cap and secure to prevent cap blocking off the flue.

4. Vent. Connector may be galvanized iron or copper. Protect all combustible materials within 6 inches of metal vent.

5. Fireproofing. Provide 3 inches clearance all around heater. Provide 1 inch clearance from class B flue, 6 inches clearance from class C flue to any combustible materials.

6. Water Connections. Install gate valve or full-way valve at cold water inlet. Install unions on hot and cold connections. Valve to control unions.

7. Gas Connection. Install approved type gas valve. Use solid connections on heaters rated in excess of 100,000 Btu.

8. Combustion Air. Provide sufficient combustion to support burner air demand.

9. Relief Valves. When such valves are required, pipe the drain pipe to outer wall and terminate the drain to a point within 24 inches on grade. Do *not* run the drain up through the roof. Trapping of relief is strictly prohibited.

10. Water Heater Safe. Install pan under heater when required. Run ample-size drain to outer wall.

11. Lights. Install electric light at heater location.

12. Installation Notes. Heater controls must be accessible at all times, with at least 18 inches clearance in front. Provide adequate support for any water heater installation. Timbers must be large enough to support the entire load that may be required. Locate heater over a hallway, if possible, so that access is kept to a minimum.

13. Do not pile combustible materials near water heater, keep access free at all times, provide adequate combustion air and ventilation.

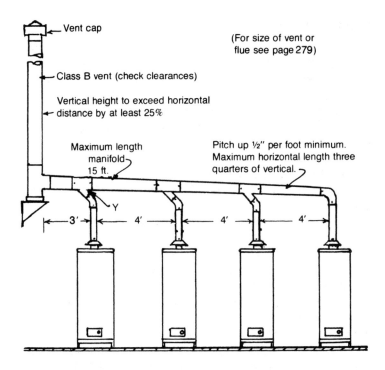

Vent cap

(For size of vent or
flue see page 279)

Class B vent (check clearances)

Vertical height to exceed horizontal
distance by at least 25%

Maximum length
manifold
15 ft.

Pitch up ½" per foot minimum.
Maximum horizontal length three
quarters of vertical.

Y

3'   4'   4'   4'

Maximum heaters on one manifold 4. Maximum distance between heaters 5 feet. All connections on Y fittings. All joints secured by bolts or metal screws. Increase manifold at each heater or
run full bore to most remote heater. Run vertically as high as
possible before entering manifold. Check all vents for clearances to
combustibles. Install gravity vents at least 12 inches above roof and
4 feet away from air intakes.

WATER HEATER IN A CLOSET

Closet Size. Closet must be at least 3 inches larger than heater on all sides. If closer than 3 inches to walls or door, provide ¼-inch asbestos board covered with 26-gauge metal as protection near burner area.

Combustion Air. A minimum of 100 square inches of ventilation must be provided to all water-heater compartments equally divided top and bottom. If a large heater is installed, the minimum requirement is 1 square inch for each 1000 Btu of burner rating equally divided top and bottom.

Vent Connector. Fireproof combustibles within 6 inches of metal vent.

Vent Cap at Roof. Vent termination must be 1 foot above roof. See that flue is not blocked by vent cap.

Water Connections. Requires fullway valve on cold water inlet. Install unions at cold water inlet and hot water outlet.

Relief Valves. If installed as a temperature relief, it must be within 3 inches of the hottest point of the heated water. Run drain to an outer wall and run to a point not more than 24 inches or less than 6 inches of outside ground, pointing downward.

Locations. No water heater may be installed in a bedroom, bathroom, or any closet off any such room or under an interior stairway.

Heater Approval. Name plate identifying heater approval, with type of gas and burner rating, shall be plainly visible. LPG burners shall be provided with not less than 100 square inches minimum air for each 1000 Btu burner rating.

Gas Shutoffs. Must be accessible, and approved type, tubing connectors may not be used where burner ratings exceed 100,000 Btu.

Vent Terminations. Minimum distance 1 foot above and 4 feet from any window, door, air intake, wall or roof angle of 45 degrees or less to the vertical. All flues shall enter Tee inlets; no class C flue may be used unless fully exposed to view in its entirety. Class B flues must have 1-inch clearance to combustible materials.

WATER HEATER IN GARAGE

1. Flue. Extend flue through roof 12 inches. See that vent cap is on and that flue is not blocked by cap.

2. Hanger or Bracket. See that it is properly fastened to support flue.

3. Connection to Flue. See that connector is of proper materials and enters flue on a Tee connection where required.

4. Fireproofing. See that all metal connectors clear combustible materials by 6 inches and all class B vents have a minimum clearance of 1 inch.

5. Water Connections. Must be joined to heater with union connections. (Check code for dielectric unions where required.)

6. Water Shutoff. Install fullway or gate valve on the cold water side, in accessible position at heater location on supply side ahead of union.

7. Gas Connection. Use rigid connection where burner rating is in excess of 100,000 Btu.

8. Gas Shutoff. Gas cock must be approved type lever handle or square head plug type without springs or as approved by the department having jurisdiction.

9. Tubing. Tubing connections may be used where burner capacity does not exceed 100,000 Btu and does not exceed 3 feet in length. No tubing through walls or in concealed locations.

10. Protection. In garages all heaters must be located where it is not subject to mechanical injury. In the case where there is a possibility of injury, suitable curbs, railings, or protection must be provided. (Check for elevation requirements of some departments.)

11. Temperature and Relief Valves. Must be located at hottest point of water heater. Where no opening is provided install relief to extend into top of heater if possible.

12. Combustion Air. Provide adequate air to burners, keep combustibles away from heater, do not install shelving or storage space at or near heater location.

WATER SOFTENER INSTALLATION

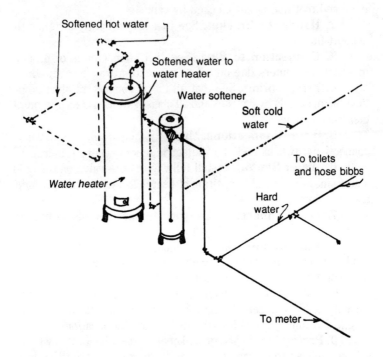

When installing water softeners, do not undersize. The inlet to the softener must be as large as the water supply.

All water softeners should be stamped with the maximum rated flow and pressure loss, and all piping to and from a water softener should be adequately sized to prevent a pressure loss.

Water softener drains are hooked up as an indirect waste and are drained to an approved plumbing fixture or air-gap fitting. When figuring water softener installations, its flow rating and pressure loss must be known.

DRAIN FROM WATER SOFTENER

Approved method of installation. Use hookup 1 and 2 where drain is automatic. Alternate method 3 may be used if the drain is manual.

Check on alternate method 3 before installing. "Fixed air gaps" are required on water softener drains to waste connection.

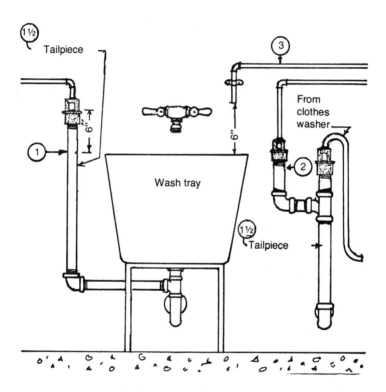

241

MINIMUM WASTE, WATER, AND TRAP SIZES FOR FIXTURES

Shower . . Minimum water hot or cold ½"
Trap size . . 2"
Vertical waste . . 2"
Vent size . . 1½"

Sink . . Minimum water hot or cold ½"
Trap size . . 1½"
Vertical waste . . 2"
Vent size . . 1½"

Tray . . Minimum water hot or cold ½"
Trap size . . 1½"
Horizontal waste . . 2"
Vertical waste . . 1½"
Vent size . . 1½"

Lavatory or wash basin
Minimum water hot or cold . . ⅜"
Horizontal waste . . 1½" (one only)
Vertical waste . . 1½" (maximum two)
Vent size . . 1¼" (vertical only)
Trap size . . 1½"

Water closet . . Minimum water ⅜"
Horizontal waste . . 4" (check for 3")
Vertical waste . . 4" (check for 3")
Vent size . . 3" (check for 4")
additional may be vented with 2"

Bath tub . . Minimum water hot or cold ½"
Horizontal waste . . 2"
Vertical waste . . 1½"
Vent size . . 1½"
Trap size . . 1½"

DIAMETER, AREA, AND CIRCUMFERENCE (CIRCLES)

Diameter	Circumference	Area	Diameter	Circumference	Area
1	3.1416	0.7854	12	37.69	113.09
1⅛	3.534	0.9940	12½	39.27	122.71
1¼	3.927	1.227	13	40.84	132.73
1⅜	4.319	1.484	13½	42.41	143.13
1½	4.712	1.767	14	43.98	153.93
1⅝	5.105	2.073	14½	45.55	165.13
1¾	5.497	2.405	15	47.12	176.71
1⅞	5.890	2.761	15½	48.69	188.69
2	6.238	3.141	16	50.26	201.06
2¼	7.068	3.976	16½	51.83	213.82
2½	7.854	4.908	17	53.40	226.98
2¾	8.639	5.939	17½	54.97	240.52
3	9.424	7.068	18	56.54	254.46
3¼	10.21	8.295	18½	58.11	268.80
3½	10.99	9.621	19	59.69	283.52
3¾	11.78	11.044	19½	61.21	298.64
4	12.56	12.566	20	62.83	314.16
4½	14.13	15.904	20½	64.40	330.06
5	15.70	19.635	21	65.97	346.36
5½	17.27	23.758	21½	67.54	363.05
6	18.84	28.274	22	69.11	380.13
6½	20.42	33.183	22½	70.68	397.60
7	21.99	38.484	23	72.25	415.47
7½	23.56	44.178	23½	73.82	433.73
8	25.13	50.265	24	74.39	452.39
8½	26.70	56.745	24½	76.96	471.43
9	28.27	63.617	25	78.54	490.87
9½	29.84	70.882	26	81.86	530.93
10	31.41	78.54	27	84.82	572.55
10½	32.98	86.59	28	87.96	615.75
11	34.55	95.03	29	91.10	660.52
11½	36.12	103.86	30	94.24	706.86

To change decimals into fractions, see fraction and decimal equivalents. For circles larger than 30 inches, see formula for circles.

GAUGE NUMBER AND THICKNESS
FOR METALS OTHER THAN IRON OR STEEL

Gauge No.	Thickness Inch	Gauge No.	Thickness Inch	Gauge No.	Thickness Inch
1	.004	13	.036	25	.095
2	.005	14	.041	26	.103
3	.008	15	.047	27	.113
4	.010	16	.051	28	.120
5	.012	17	.057	29	.124
6	.013	18	.061	30	.126
7	.015	19	.064	31	.133
8	.016	20	.067	32	.143
9	.019	21	.072	33	.145
10	.024	22	.074	34	.148
11	.029	23	.077	35	.158
12	.034	24	.082	36	.167

BOILING POINTS OF VARIOUS LIQUIDS AND FLUIDS

Degrees Fahrenheit

Water, Atmospheric Pressure	212
Alcohol	173
Turpentine	315
Ammonia	140
Linseed oil	597
Sulphuric Acid	590
Sulphur	833
Nitric Acid	248
Ether, Sulphuric	100
Saturated Brine	226
Mercury	676

MELTING POINTS OF DIFFERENT METALS

Metal	Degrees Fahrenheit
Aluminum	1214
Antimony	1169
Bismuth	507
Brass	1900
Bronze	1692
Copper	1943
Gold (pure)	1947
Iron (cast, gray)	2228
Iron (cast, white)	2075
Iron (wrought)	2737
Lead	622
Platinum	3110
Silver (pure)	1751
Steel	2500
Tin	449
Zinc	786

BOILING POINTS OF WATER AT VARIOUS ALTITUDES

	Degrees Fahrenheit
Sea level	212
512 Elevation	211
1025 Elevation	210
2063 Elevation	208
4169 Elevation	204
6304 Elevation	200
7932 Elevation	197
9031 Elevation	195
10127 Elevation	193
11799 Elevation	190
12934 Elevation	188
14075 Elevation	186
15221 Elevation	184

Water boils in a direct proportion to the atmospheric pressure and the point or elevation above sea level.

FRICTION IN PIPE BENDS

Diameter of Bend in inches ↓	Friction in the bend is to friction in ft. of pipe ↓	Friction in bend corresponding to straight pipe by diameters of Fittings
5″ . . . . 20′	. . . . 48 dia. of.	
4″ . . . . 15′	. . . . 45 diameters	
3″ . . . . 9′	. . . . 36 diameters	
2″ . . . . 5′	. . . . 30 diameters	
1½″ . . . . 3′3″	. . . . 27 diameters	
1¼″ . . . . 2′6″	. . . . . . 24 diameters	
1 . . . . 1′6″	. . . . 18 diameters	

2″ pipe for example would take 5 feet or 60 inches to equal the friction of one 2″ bend

FRICTION OF FITTINGS

Kind of Fitting	Number of 90° bends it is equal to in frictional resistance
Coupling	one-tenth of 90° bend Twice that of the same length of straight pipe
45°	½ of a 90° bend
Return bend (open	Same as 90° bend
Tee Fitting	Equals 2 . . . 90° bends
Gate Valve	½ of a 90° bend
Globe Valve	12 times that of a 90° bend

When figuring the capacity of gas or water pipe always allow for fitting loss in addition to pipe loss.

EXPANSION IN PIPE LINES

Inches of linear expansion per 100 feet. Table in 20-degree variables of Fahrenheit scale.

The expansion for any length of pipe may be determined by the following method. From the table, check the difference in increased length at the minimum and maximum temperature anticipated. Divide this result by 100 to obtain the increase in length per foot. Once this length per foot is determined, multiply this by the length of line in feet.

Temp. Degrees F	Cast iron	Steel	Wrought iron	Copper	Brass	A.B.S plastic
	inches per 100 feet					
0°	.0	0	0	0	0	0
20°	.128	.15	.155	.25	.25	1.34
40°	.263	.30	.31	.45	.45	2.68
60°	.391	.455	.475	.65	.67	4.02
80°	.522	.61	.63	.87	.92	5.36
100°	.66	.77	.80	1.10	1.17	6.70
120°	.799	.915	.96	1.35	1.41	8.04
140°	.924	1.075	1.13	1.57	1.69	9.38
160°	1.073	1.235	1.29	1.77	1.91	10.72
180°	1.218	1.40	1.46	2.00	2.15	12.06
200°	1.368	1.57	1.64	2.25	2.43	13.40
220°		1.73	1.81	2.52	2.71	14.74
240°		1.89	1.98	2.74	2.97	
260°		2.065	2.16	2.95	3.22	
280°		2.23	2.335	3.17	3.47	
300°		2.41	2.52	3.42	3.76	
320°		2.59	2.70	3.70	4.07	
340°		2.76	2.87	3.95	4.30	
360°		2.935	3.05	4.15	4.58	
380°		3.11	3.235	4.40	4.85	
400°		3.29	3.43	4.64	5.14	
420°		3.465	3.62	4.89	5.45	
440°		3.65	3.805	5.15	5.73	
460°		3.835	4.00	5.37	5.96	
480°		4.02	4.19	5.64	6.25	
500°		4.21	4.39	5.88	6.55	

FLOW CAPACITIES OF CAST-IRON DRAINS

Size of Pipe	Pitch per Ft.	Gallons Per Min.	Velocity Ft. Per Sec.
2"	⅛"	12.6	1.28
2"	¼"	17.6	1.80
2"	½"	27.8	2.85
2"	1"	40.	4.00
3"	⅛"	36.2	1.65
3"	¼"	51.3	2.34
3"	½"	80.3	3.70
3"	1"	113.8	5.24
4"	⅛"	77.4	1.97
4"	¼"	109.6	2.79
4"	½"	173.6	4.42
4"	1"	245.4	6.25
5"	⅛"	138.6	2.25
5"	¼"	196.5	3.19
5"	½"	311.0	5.05
5"	1"	439.2	7.13
6"	⅛"	221.4	2.51
6"	¼"	313.1	3.55
6"	½"	494.8	5.61
6"	1"	699.4	7.93
8"	⅛"	463.0	2.94
8"	¼"	656.0	4.16
8"	½"	1037.0	6.58
8"	1"	1468.0	9.31
10"	⅛"	814.0	3.33
10"	¼"	1149.0	4.70
10"	½"	1819.0	7.44
10"	1"	2572.0	10.52
12"	⅛"	1300.0	3.68
12"	¼"	1838.0	5.20
12"	½"	2905.0	8.22
12"	1"	4034.0	11.62

The above flow capacities may be used to figure a safe capacity of old lines that may be lined with deposit and in which there is no static head.

TABLES OF SQUARES, CUBES, SQUARE ROOTS, AND CUBE ROOTS OF NUMBERS FROM 1 TO 1000

No.	Square	Cube	Square Root	Cube Root
1	1	1	1.0000	1.0000
2	4	8	1.4142	1.2599
3	9	27	1.7321	1.4422
4	16	64	2.0000	1.5874
5	25	125	2.2361	1.7100
6	36	216	2.4495	1.8171
7	49	343	2.6458	1.9129
8	64	512	2.8284	2.0000
9	81	729	3.0000	2.0801
10	100	1000	3.1623	2.1544
11	121	1331	3.3166	2.2240
12	144	1728	3.4641	2.2894
13	169	2197	3.6056	2.3513
14	196	2744	3.7417	2.4101
15	225	3375	3.8730	2.4662
16	256	4096	4.0000	2.5198
17	289	4913	4.1231	2.5713
18	324	5832	4.2426	2.6207
19	361	6859	4.3589	2.6684
20	400	8000	4.4721	2.7144
21	441	9261	4.5826	2.7589
22	484	10648	4.6904	2.8020
23	529	12167	4.7958	2.8439
24	576	13824	4.8990	2.8845
25	625	15625	5.0000	2.9240
26	676	17576	5.0990	2.9625
27	729	19683	5.1962	3.0000
28	784	21952	5.2915	3.0366
29	841	24389	5.3852	3.0723
30	900	27000	5.4772	3.1072
31	961	29791	5.5678	3.1414
32	1024	32768	5.6569	3.1748
33	1089	35937	5.7446	3.2075
34	1156	39304	5.8310	3.2396
35	1225	42875	5.9161	3.2711
36	1296	46656	6.0000	3.3019
37	1369	50653	6.0828	3.3322
38	1444	54872	6.1644	3.3620
39	1521	59319	6.2450	3.3912
40	1600	64000	6.3246	3.4200

No.	Square	Cube	Square Root	Cube Root
41	1681	68921	6.4031	3.4482
42	1764	74088	6.4807	3.4760
43	1849	79507	6.5574	3.5034
44	1936	85184	6.6332	3.5303
45	2025	91125	6.7082	3.5569
46	2116	97336	6.7823	3.5830
47	2209	103823	6.8557	3.6088
48	2304	110592	6.9282	3.6342
49	2401	117649	7.0000	3.6593
50	2500	125000	7.0711	3.6840
51	2601	132651	7.1414	3.7084
52	2704	140608	7.2111	3.7325
53	2809	148877	7.2801	3.7563
54	2916	157464	7.3485	3.7798
55	3025	166375	7.4162	3.8030
56	3136	175616	7.4833	3.8259
57	3249	185193	7.5498	3.8485
58	3364	195112	7.6158	3.8709
59	3481	205379	7.6811	3.8930
60	3600	216000	7.7460	3.9149
61	3721	226981	7.8102	3.9365
62	3844	238328	7.8740	3.9579
63	3969	250047	7.9373	3.9791
64	4096	262144	8.0000	4.
65	4225	274625	8.0623	4.0207
66	4356	287496	8.1240	4.0412
67	4489	300763	8.1854	4.0615
68	4624	314432	8.2462	4.0817
69	4761	328509	8.3066	4.1016
70	4900	343000	8.3666	4.1213
71	5041	357911	8.4261	4.1408
72	5184	373248	8.4853	4.1602
73	5329	389017	8.5440	4.1793
74	5476	405224	8.6023	4.1983
75	5625	421875	8.6603	4.2172
76	5776	438976	8.7178	4.2358
77	5929	456533	8.7750	4.2543
78	6084	474552	8.8318	4.2727
79	6241	493039	8.8882	4.2908
80	6400	512000	8.9443	4.3089
81	6561	531441	9.	4.3267
82	6724	551368	9.0554	4.3445
83	6889	571787	9.1104	4.3621
84	7056	592704	9.1652	4.3795
85	7225	614125	9.2195	4.3968

No.	Square	Cube	Square Root	Cube Root
86	7396	636056	9.2736	4.4140
87	7569	658503	9.3274	4.4310
88	7744	681472	9.3808	4.4480
89	7921	704969	9.4340	4.4647
90	8100	729000	9.4868	4.4814
91	8281	753571	9.5394	4.4979
92	8464	778688	9.5917	4.5144
93	8649	804357	9.6437	4.5307
94	8836	830584	9.6954	4.5468
95	9025	857375	9.7468	4.5629
96	9216	884736	9.7980	4.5789
97	9409	912673	9.8489	4.5947
98	9604	941192	9.8995	4.6104
99	9801	970299	9.9499	4.6261
100	10000	1000000	10.	4.6416
101	10201	1030301	10.0499	4.6570
102	10404	1061208	10.0995	4.6723
103	10609	1092727	10.1489	4.6875
104	10816	1124864	10.1980	4.7027
105	11025	1157625	10.2470	4.7177
106	11236	1191016	10.2956	4.7326
107	11449	1225043	10.3441	4.7475
108	11664	1259712	10.3923	4.7622
109	11881	1295029	10.4403	4.7769
110	12100	1331000	10.4881	4.7914
111	12321	1367631	10.5357	4.8059
112	12544	1404928	10.5830	4.8203
113	12769	1442897	10.6301	4.8346
114	12996	1481544	10.6771	4.8488
115	13225	1520875	10.7238	4.8629
116	13456	1560896	10.7703	4.8770
117	13689	1601613	10.8167	4.8910
118	13924	1643032	10.8628	4.9049
119	14161	1685159	10.9087	4.9187
120	14400	1728000	10.9545	4.9324
121	14641	1771561	11.	4.9461
122	14884	1815848	11.0454	4.9597
123	15129	1860867	11.0905	4.9732
124	15376	1906624	11.1355	4.9866
125	15625	1953125	11.1803	5.
126	15876	2000376	11.2250	5.0133
127	16129	2048383	11.2694	5.0265
128	16384	2097152	11.3137	5.0397
129	16641	2146689	11.3578	5.0528
130	16900	2197000	11.4018	5.0658

No.	Square	Cube	Square Root	Cube Root
131	17161	2248091	11.4455	5.0788
132	17424	2299968	11.4891	5.0916
133	17689	2352637	11.5326	5.1045
134	17956	2406104	11.5758	5.1172
135	18225	2460375	11.6190	5.1299
136	18496	2515456	11.6619	5.1426
137	18769	2571353	11.7047	5.1551
138	19044	2628072	11.7473	5.1676
139	19321	2685619	11.7898	5.1801
140	19600	2744000	11.8322	5.1925
141	19881	2803221	11.8743	5.2048
142	20164	2863288	11.9164	5.2171
143	20449	2924207	11.9583	5.2293
144	20736	2985984	12.	5.2415
145	21025	3048625	12.0416	5.2536
146	21316	3112136	12.0830	5.2656
147	21609	3176523	12.1244	5.2776
148	21904	3241792	12.1655	5.2896
149	22201	3307949	12.2066	5.3015
150	22500	3375000	12.2474	5.3133
151	22801	3442951	12.2882	5.3251
152	23104	3511808	12.3288	5.3368
153	23409	3581577	12.3693	5.3485
154	23716	3652264	12.4097	5.3601
155	24025	3723875	12.4499	5.3717
156	24336	3796416	12.4900	5.3832
157	24649	3869893	12.5300	5.3947
158	24964	3944312	12.5698	5.4061
159	25281	4019679	12.6095	5.4175
160	25600	4096000	12.6491	5.4288
161	25921	4173281	12.6886	5.4401
162	26244	4251528	12.7279	5.4514
163	26569	4330747	12.7671	5.4626
164	26896	4410944	12.8062	5.4737
165	27225	4492125	12.8452	5.4848
166	27556	4574296	12.8841	5.4959
167	27889	4657463	12.9228	5.5069
168	28224	4741632	12.9615	5.5178
169	28561	4826809	13.	5.5288
170	28900	4913000	13.0384	5.5397
171	29241	5000211	13.0767	5.5505
172	29584	5088448	13.1149	5.5613
173	29929	5177717	13.1529	5.5721
174	30276	5268024	13.1909	5.5828
175	30625	5359375	13.2288	5.5934

No.	Square	Cube	Square Root	Cube Root
176	30976	5451776	13.2665	5.6041
177	31329	5545233	13.3041	5.6147
178	31684	5639752	13.3417	5.6252
179	32041	5735339	13.3791	5.6357
180	32400	5832000	13.4164	5.6462
181	32761	5929741	13.4536	5.6567
182	33124	6028568	13.4907	5.6671
183	33489	6128487	13.5277	5.6774
184	33856	6229504	13.5647	5.6877
185	34225	6331625	13.6015	5.6980
186	34596	6434856	13.6382	5.7083
187	34969	6539203	13.6748	5.7185
188	35344	6644672	13.7113	5.7287
189	35721	6751269	13.7477	5.7388
190	36100	6859000	13.7840	5.7489
191	36481	6967871	13.8203	5.7590
192	36864	7077888	13.8564	5.7690
193	37249	7189057	13.8924	5.7790
194	37636	7301384	13.9284	5.7890
195	38025	7414875	13.9642	5.7989
196	38416	7529536	14.	5.8088
197	38809	7645373	14.0357	5.8186
198	39204	7762392	14.0712	5.8285
199	39601	7880599	14.1067	5.8383
200	40000	8000000	14.1421	5.8480
201	40401	8120601	14.1774	5.8578
202	40804	8242408	14.2127	5.8675
203	41209	8365427	14.2478	5.8771
204	41616	8489664	14.2829	5.8868
205	42025	8615125	14.3178	5.8964
206	42436	8741816	14.3527	5.9059
207	42849	8869743	14.3875	5.9155
208	43264	8998912	14.4222	5.9250
209	43681	9129329	14.4568	5.9345
210	44100	9261000	14.4914	5.9439
211	44521	9393931	14.5258	5.9533
212	44944	9528128	14.5602	5.9627
213	45369	9663597	14.5945	5.9721
214	45796	9800344	14.6287	5.9814
215	46225	9938375	14.6629	5.9907

No.	Square	Cube	Square Root	Cube Root
216	46656	10077696	14.6969	6.
217	47089	10218313	14.7309	6.0092
218	47524	10360232	14.7648	6.0185
219	47961	10503459	14.7986	6.0277
220	48400	10648000	14.8324	6.0368
221	48841	10793861	14.8661	6.0459
222	49284	10941048	14.8997	6.0550
223	49729	11089567	14.9332	6.0641
224	50176	11239424	14.9666	6.0732
225	50625	11390625	15.	6.0822
226	51076	11543176	15.0333	6.0912
227	51529	11697083	15.0665	6.1002
228	51984	11852352	15.0997	6.1091
229	52441	12008989	15.1327	6.1180
230	52900	12167000	15.1658	6.1269
231	53361	12326391	15.1987	6.1358
232	53824	12487168	15.2315	6.1446
233	54289	12649337	15.2643	6.1534
234	54756	12812904	15.2971	6.1622
235	55225	12977875	15.3297	6.1710
236	55696	13144256	15.3623	6.1797
237	56169	13312053	15.3948	6.1885
238	56644	13481272	15.4272	6.1972
239	57121	13651919	15.4596	6.2058
240	57600	13824000	15.4919	6.2145
241	58081	13997521	15.5242	6.2231
242	58564	14172488	15.5563	6.2317
243	59049	14348907	15.5885	6.2403
244	59536	14526784	15.6205	6.2488
245	60025	14706125	15.6525	6.2573
246	60516	14886936	15.6844	6.2658
247	61009	15069223	15.7162	6.2743
248	61504	15252992	15.7480	6.2828
249	62001	15438249	15.7797	6.2912
250	62500	15625000	15.8114	6.2996
251	63001	15813251	15.8430	6.3080
252	63504	16003008	15.8745	6.3164
253	64009	16194277	15.9060	6.3247
254	64516	16387064	15.9374	6.3330
255	65025	16581375	15.9687	6.3413
256	65536	16777216	16.	6.3496
257	66049	16974593	16.0312	6.3579
258	66564	17173512	16.0624	6.3661
259	67081	17373979	16.0935	6.3743
260	67600	17576000	16.1245	6.3825

No.	Square	Cube	Square Root	Cube Root
261	68121	17779581	16.1555	6.3907
262	68644	17984728	16.1864	6.3988
263	69169	18191447	16.2173	6.4070
264	69696	18399744	16.2481	6.4151
265	70225	18609625	16.2788	6.4232
266	70756	18821096	16.3095	6.4312
267	71289	19034163	16.3401	6.4393
268	71824	19248832	16.3707	6.4473
269	72361	19465109	16.4012	6.4553
270	72900	19683000	16.4317	6.4633
271	73441	19902511	16.4621	6.4713
272	73984	20123648	16.4924	6.4792
273	74529	20346417	16.5227	6.4872
274	75076	20570824	16.5529	6.4951
275	75625	20796875	16.5831	6.5030
276	76176	21024576	16.6132	6.5108
277	76729	21253933	16.6433	6.5187
278	77284	21484952	16.6733	6.5265
279	77841	21717639	16.7033	6.5343
280	78400	21952000	16.7332	6.5421
281	78961	22188041	16.7631	6.5499
282	79524	22425768	16.7929	6.5577
283	80089	22665187	16.8226	6.5654
284	80656	22906304	16.8523	6.5731
285	81225	23149125	16.8819	6.5808
286	81796	23393656	16.9115	6.5885
287	82369	23639903	16.9411	6.5962
288	82944	23887872	16.9706	6.6039
289	83521	24137569	17.	6.6115
290	84100	24389000	17.0294	6.6191
291	84681	24642171	17.0587	6.6267
292	85264	24897088	17.0880	6.6343
293	85849	25153757	17.1172	6.6419
294	86436	25412184	17.1464	6.6494
295	87025	25672375	17.1756	6.6569
296	87616	25934336	17.2047	6.6644
297	88209	26198073	17.2337	6.6719
298	88804	26463592	17.2627	6.6794
299	89401	26730899	17.2916	6.6869
300	90000	27000000	17.3205	6.6943
301	90601	27270901	17.3494	6.7018
302	91204	27543608	17.3781	6.7092
303	91809	27818127	17.4069	6.7166
304	92416	28094464	17.4356	6.7240
305	93025	28372625	17.4642	6.7313

No.	Square	Cube	Square Root	Cube Root
306	93636	28652616	17.4929	6.7387
307	94249	28934443	17.5214	6.7460
308	94864	29218112	17.5499	6.7533
309	95481	29503629	17.5784	6.7606
310	96100	29791000	17.6068	6.7679
311	96721	30080231	17.6352	6.7752
312	97344	30371328	17.6635	6.7824
313	97969	30664297	17.6918	6.7897
314	98596	30959144	17.7200	6.7969
315	99225	31255875	17.7482	6.8041
316	99856	31554496	17.7764	6.8113
317	100489	31855013	17.8045	6.8185
318	101124	32157432	17.8326	6.8256
319	101761	32461759	17.8606	6.8328
320	102400	32768000	17.8885	6.8399
321	103041	33076161	17.9165	6.8470
322	103684	33386248	17.9444	6.8541
323	104329	33698267	17.9722	6.8612
324	104976	34012224	18.	6.8683
325	105625	34328125	18.0278	6.8753
326	106276	34645976	18.0555	6.8824
327	106929	34965783	18.0831	6.8894
328	107584	35287552	18.1108	6.8964
329	108241	35611289	18.1384	6.9034
330	108900	35937000	18.1659	6.9104
331	109561	36264691	18.1934	6.9174
332	110224	36594368	18.2209	6.9244
333	110889	36926037	18.2483	6.9313
334	111556	37259704	18.2757	6.9382
335	112225	37595375	18.3030	6.9451
336	112896	37933056	18.3303	6.9521
337	113569	38272753	18.3576	6.9589
338	114244	38614472	18.3848	6.9658
339	114921	38958219	18.4120	6.9727
340	115600	39304000	18.4391	6.9795
341	116281	39651821	18.4662	6.9864
342	116964	40001688	18.4932	6.9932
343	117649	40353607	18.5203	7.
344	118336	40707584	18.5472	7.0068
345	119025	41063625	18.5742	7.0136
346	119716	41421736	18.6011	7.0203
347	120409	41781923	18.6279	7.0271
348	121104	42144192	18.6548	7.0338
349	121801	42508549	18.6815	7.0406
350	122500	42875000	18.7083	7.0473

No.	Square	Cube	Square Root	Cube Root
351	123201	43243551	18.7350	7.0540
352	123904	43614208	18.7617	7.0607
353	124609	43986977	18.7883	7.0674
354	125316	44361864	18.8149	7.0740
355	126025	44738875	18.8414	7.0807
356	126736	45118016	18.8680	7.0873
357	127449	45499293	18.8944	7.0940
358	128164	45882712	18.9209	7.1006
359	128881	46268279	18.9473	7.1072
360	129600	46656000	18.9737	7.1138
361	130321	47045881	19.	7.1204
362	131044	47437928	19.0263	7.1269
363	131769	47832147	19.0526	7.1335
364	132496	48228544	19.0788	7.1400
365	133225	48627125	19.1050	7.1466
366	133956	49027896	19.1311	7.1531
367	134689	49430863	19.1572	7.1596
368	135424	49836032	19.1833	7.1661
369	136161	50243409	19.2094	7.1726
370	136900	50653000	19.2354	7.1791
371	137641	51064811	19.2614	7.1855
372	138384	51478848	19.2873	7.1920
373	139129	51895117	19.3132	7.1984
374	139876	52313624	19.3391	7.2048
375	140625	52734375	19.3649	7.2112
376	141376	53157376	19.3907	7.2177
377	142129	53582633	19.4165	7.2240
378	142884	54010152	19.4422	7.2304
379	143641	54439939	19.4679	7.2368
380	144400	54872000	19.4936	7.2432
381	145161	55306341	19.5192	7.2495
382	145924	55742968	19.5448	7.2558
383	146689	56181887	19.5704	7.2622
384	147456	56623104	19.5959	7.2685
385	148225	57066625	19.6214	7.2748
386	148996	57512456	19.6469	7.2811
387	149769	57960603	19.6723	7.2874
388	150544	58411072	19.6977	7.2936
389	151321	58863869	19.7231	7.2999
390	152100	59319000	19.7484	7.3061
391	152881	59776471	19.7737	7.3124
392	153664	60236288	19.7990	7.3186
393	154449	60698457	19.8242	7.3248
394	155236	61162984	19.8494	7.3310
395	156025	61629875	19.8746	7.3372

No.	Square	Cube	Square Root	Cube Root
396	156816	62099136	19.8997	7.3434
397	157609	62570773	19.9249	7.3496
398	158404	63044792	19.9499	7.3558
399	159201	63521199	19.9750	7.3619
400	160000	64000000	20.	7.3681
401	160801	64481201	20.0250	7.3742
402	161604	64964808	20.0499	7.3803
403	162409	65450827	20.0749	7.3864
404	163216	65939264	20.0998	7.3925
405	164025	66430125	20.1246	7.3986
406	164836	66923416	20.1494	7.4047
407	165649	67419143	20.1742	7.4108
408	166464	67917312	20.1990	7.4169
409	167281	68417929	20.2237	7.4229
410	168100	68921000	20.2485	7.4290
411	168921	69426531	20.2731	7.4350
412	169744	69934528	20.2978	7.4410
413	170569	70444997	20.3224	7.4470
414	171396	70957944	20.3470	7.4530
415	172225	71473375	20.3715	7.4590
416	173056	71991296	20.3961	7.4650
417	173889	72511713	20.4206	7.4710
418	174724	73034632	20.4450	7.4770
419	175561	73560059	20.4695	7.4829
420	176400	74088000	20.4939	7.4889
421	177241	74618461	20.5183	7.4948
422	178084	75151448	20.5426	7.5007
423	178929	75686967	20.5670	7.5067
424	179776	76225024	20.5913	7.5126
425	180625	76765625	20.6155	7.5185
426	181476	77308776	20.6398	7.5244
427	182329	77854483	20.6640	7.5302
428	183184	78402752	20.6882	7.5361
429	184041	78953589	20.7123	7.5420
430	184900	79507000	20.7364	7.5478
431	185761	80062991	20.7605	7.5537
432	186624	80621568	20.7846	7.5595
433	187489	81182737	20.8087	7.5654
434	188356	81746504	20.8327	7.5712
435	189225	82312875	20.8567	7.5770
436	190096	82881856	20.8806	7.5828
437	190969	83453453	20.9045	7.5886
438	191844	84027672	20.9284	7.5944
439	192721	84604519	20.9523	7.6001
440	193600	85184000	20.9762	7.6059

No.	Square	Cube	Square Root	Cube Root
441	194481	85766121	21.	7.6117
442	195364	86350888	21.0238	7.6174
443	196249	86938307	21.0476	7.6232
444	197136	87528384	21.0713	7.6289
445	198025	88121125	21.0950	7.6346
446	198916	88716536	21.1187	7.6403
447	199809	89314623	21.1424	7.6460
448	200704	89915392	21.1660	7.6517
449	201601	90518849	21.1896	7.6574
450	202500	91125000	21.2132	7.6631
451	203401	91733851	21.2368	7.6688
452	204304	92345408	21.2603	7.6744
453	205209	92959677	21.2838	7.6801
454	206116	93576664	21.3073	7.6857
455	207025	94196375	21.3307	7.6914
456	207936	94818816	21.3542	7.6970
457	208849	95443993	21.3776	7.7026
458	209764	96071912	21.4009	7.7082
459	210681	96702579	21.4243	7.7138
460	211600	97336000	21.4476	7.7194
461	212521	97972181	21.4709	7.7250
462	213444	98611128	21.4942	7.7306
463	214369	99252847	21.5174	7.7362
464	215296	99897344	21.5407	7.7418
465	216225	100544625	21.5639	7.7473
466	217156	101194696	21.5870	7.7529
467	218089	101847563	21.6102	7.7584
468	219024	102503232	21.6333	7.7639
469	219961	103161709	21.6564	7.7695
470	220900	103823000	21.6795	7.7750
471	221841	104487111	21.7025	7.7805
472	222784	105154048	21.7256	7.7860
473	223729	105823817	21.7486	7.7915
474	224676	106496424	21.7715	7.7970
475	225625	107171875	21.7945	7.8025
476	226576	107850176	21.8174	7.8079
477	227529	108531333	21.8403	7.8134
478	228484	109215352	21.8632	7.8188
479	229441	109902239	21.8861	7.8243
480	230400	110592000	21.9089	7.8297
481	231361	111284641	21.9317	7.8352
482	232324	111980168	21.9545	7.8406
483	233289	112678587	21.9773	7.8460
484	234256	113379904	22.	7.8514
485	235225	114084125	22.0227	7.8568

No.	Square	Cube	Square Root	Cube Root
486	236196	114791256	22.0454	7.8622
487	237169	115501303	22.0681	7.8676
488	238144	116214272	22.0907	7.8730
489	239121	116930169	22.1133	7.8784
490	240100	117649000	22.1359	7.8837
491	241081	118370771	22.1585	7.8891
492	242064	119095488	22.1811	7.8944
493	243049	119823157	22.2036	7.8998
494	244036	120553784	22.2261	7.9051
495	245025	121287375	22.2486	7.9105
496	246016	122023936	22.2711	7.9158
497	247009	122763473	22.2935	7.9211
498	248004	123505992	22.3159	7.9264
499	249001	124251499	22.3383	7.9317
500	250000	125000000	22.3607	7.9370
501	251001	125751501	22.3830	7.9423
502	252004	126506008	22.4054	7.9476
503	253009	127263527	22.4277	7.9528
504	254016	128024064	22.4499	7.9581
505	255025	128787625	22.4722	7.9634
506	256036	129554216	22.4944	7.9686
507	257049	130323843	22.5167	7.9739
508	258064	131096512	22.5389	7.9791
509	259081	131872229	22.5610	7.9843
510	260100	132651000	22.5832	7.9896
511	261121	133432831	22.6053	7.9948
512	262144	134217728	22.6274	8.
513	263169	135005697	22.6495	8.0052
514	264196	135796744	22.6716	8.0104
515	265225	136590875	22.6936	8.0156
516	266256	137388096	22.7156	8.0208
517	267289	138188413	22.7376	8.0260
518	268324	138991832	22.7596	8.0311
519	269361	139798359	22.7816	8.0363
520	270400	140608000	22.8035	8.0415
521	271441	141420761	22.8254	8.0466
522	272484	142236648	22.8473	8.0517
523	273529	143055667	22.8692	8.0569
524	274576	143877824	22.8910	8.0620
525	275625	144703125	22.9129	8.0671
526	276676	145531576	22.9347	8.0723
527	277729	146363183	22.9565	8.0774
528	278784	147197952	22.9783	8.0825
529	279841	148035889	23.	8.0876
530	280900	148877000	23.0217	8.0927

No.	Square	Cube	Square Root	Cube Root
531	281961	149721291	23.0434	8.0978
532	283024	150568768	23.0651	8.1028
533	284089	151419437	23.0868	8.1079
534	285156	152273304	23.1084	8.1130
535	286225	153130375	23.1301	8.1180
536	287296	153990656	23.1517	8.1231
537	288369	154854153	23.1733	8.1281
538	289444	155720872	23.1948	8.1332
539	290521	156590819	23.2164	8.1382
540	291600	157464000	23.2379	8.1433
541	292681	158340421	23.2594	8.1483
542	293764	159220088	23.2809	8.1533
543	294849	160103007	23.3024	8.1583
544	295936	160989184	23.3238	8.1633
545	297025	161878625	23.3452	8.1683
546	298116	162771336	23.3666	8.1733
547	299209	163667323	23.3880	8.1783
548	300304	164566592	23.4094	8.1833
549	301401	165469149	23.4307	8.1882
550	302500	166375000	23.4521	8.1932
551	303601	167284151	23.4734	8.1982
552	304704	168196608	23.4947	8.2031
553	305809	169112377	23.5160	8.2081
554	306916	170031464	23.5372	8.2130
555	308025	170953875	23.5584	8.2180
556	309136	171879616	23.5797	8.2229
557	310249	172808693	23.6008	8.2278
558	311364	173741112	23.6220	8.2327
559	312481	174676879	23.6432	8.2377
560	313600	175616000	23.6643	8.2426
561	314721	176558481	23.6854	8.2475
562	315844	177504328	23.7065	8.2524
563	316969	178453547	23.7276	8.2573
564	318096	179406144	23.7487	8.2621
565	319225	180362125	23.7697	8.2670
566	320356	181321496	23.7908	8.2719
567	321489	182284263	23.8118	8.2768
568	322624	183250432	23.8328	8.2816
569	323761	184220009	23.8537	8.2865
570	324900	185193000	23.8747	8.2913
571	326041	186169411	23.8956	8.2962
572	327184	187149248	23.9165	8.3010
573	328329	188132517	23.9374	8.3059
574	329476	189119224	23.9583	8.3107
575	330625	190109375	23.9792	8.3155

No.	Square	Cube	Square Root	Cube Root
576	331776	191102976	24.	8.3203
577	332929	192100033	24.0208	8.3251
578	334084	193100552	24.0416	8.3300
579	335241	194104539	24.0624	8.3348
580	336400	195112000	24.0832	8.3396
581	337561	196122941	24.1039	8.3443
582	338724	197137368	24.1247	8.3491
583	339889	198155287	24.1454	8.3539
584	341056	199176704	24.1661	8.3587
585	342225	200201625	24.1868	8.3634
586	343396	201230056	24.2074	8.3682
587	344569	202262003	24.2281	8.3730
588	345744	203297472	24.2487	8.3777
589	346921	204336469	24.2693	8.3825
590	348100	205379000	24.2899	8.3872
591	349281	206425071	24.3105	8.3919
592	350464	207474688	24.3311	8.3967
593	351649	208527857	24.3516	8.4014
594	352836	209584584	24.3721	8.4061
595	354025	210644875	24.3926	8.4108
596	355216	211708736	24.4131	8.4155
597	356409	212776173	24.4336	8.4202
598	357604	213847192	24.4540	8.4249
599	358801	214921799	24.4745	8.4296
600	360000	216000000	24.4949	8.4343
601	361201	217081801	24.5153	8.4390
602	362404	218167208	24.5357	8.4437
603	363609	219256227	24.5561	8.4484
604	364816	220348864	24.5764	8.4530
605	366025	221445125	24.5967	8.4577
606	367236	222545016	24.6171	8.4623
607	368449	223648543	24.6374	8.4670
608	369664	224755712	24.6577	8.4716
609	370881	225866529	24.6779	8.4763
610	372100	226981000	24.6982	8.4809
611	373321	228099131	24.7184	8.4856
612	374544	229220928	24.7386	8.4902
613	375769	230346397	24.7588	8.4948
614	376996	231475544	24.7790	8.4994
615	378225	232608375	24.7992	8.5040
616	379456	233744896	24.8193	8.5086
617	380689	234885113	24.8395	8.5132
618	381924	236029032	24.8596	8.5178
619	383161	237176659	24.8797	8.5224
620	384400	238328000	24.8998	8.5270

262

No.	Square	Cube	Square Root	Cube Root
621	385641	239483061	24.9199	8.5316
622	386884	240641848	24.9399	8.5362
623	388129	241804367	24.9600	8.5408
624	389376	242970624	24.9800	8.5453
625	390625	244140625	25.	8.5499
626	391876	245314376	25.0200	8.5544
627	393129	246491883	25.0400	8.5590
628	394384	247673152	25.0599	8.5635
629	395641	248858189	25.0799	8.5681
630	396900	250047000	25.0998	8.5726
631	398161	251239591	25.1197	8.5772
632	399424	252435968	25.1396	8.5817
633	400689	253636137	25.1595	8.5862
634	401956	254840104	25.1794	8.5907
635	403225	256047875	25.1992	8.5952
636	404496	257259456	25.2190	8.5997
637	405769	258474853	25.2389	8.6043
638	407044	259694072	25.2587	8.6088
639	408321	160917119	25.2784	8.6132
640	409600	262144000	25.2982	8.6177
641	410881	263374721	25.3180	8.6222
642	412164	264609288	25.3377	8.6267
643	413449	265847707	25.3574	8.6312
644	414736	267089984	25.3772	8.6357
645	416025	268336125	25.3969	8.6401
646	417316	269586136	25.4165	8.6446
647	418609	270840023	25.4362	8.6490
648	419904	272097792	25.4558	8.6535
649	421201	273359449	25.4755	8.6579
650	422500	274625000	25.4951	8.6624
651	423801	275894451	25.5147	8.6668
652	425104	277167808	25.5343	8.6713
653	426409	278445077	25.5539	8.6757
654	427716	279726264	25.5734	8.6801
655	429025	281011375	25.5930	8.6845
656	430336	282300416	25.6125	8.6890
657	431649	283593393	25.6320	8.6934
658	432964	284890312	25.6515	8.6978
659	434281	286191179	25.6710	8.7022
660	435600	287496000	25.6905	8.7066
661	436921	288804781	25.7099	8.7110
662	438244	290117528	25.7294	8.7154
663	439569	291434247	25.7488	8.7198
664	440896	292754944	25.7682	8.7241
665	442225	294079625	25.7876	8.7285

No.	Square	Cube	Square Root	Cube Root
666	443556	295408296	25.8070	8.7329
667	444889	296740963	25.8263	8.7373
668	446224	298077632	25.8457	8.7416
669	447561	299418309	25.8650	8.7460
670	448900	300763000	25.8844	8.7503
671	450241	302111711	25.9037	8.7547
672	451584	303464448	25.9230	8.7590
673	452929	304821217	25.9422	8.7634
674	454276	306182024	25.9615	8.7677
675	455625	307546875	25.9808	8.7721
676	456976	308915776	26.	8.7764
677	458329	310288733	26.0192	8.7807
678	459684	311665752	26.0384	8.7850
679	461041	313046839	26.0576	8.7893
680	462400	314432000	26.0768	8.7937
681	463761	315821241	26.0960	8.7980
682	465124	317214568	26.1151	8.8023
683	466489	318611987	26.1343	8.8066
684	467856	320013504	26.1534	8.8109
685	469225	321419125	26.1725	8.8152
686	470596	322828856	26.1916	8.8194
687	471969	324242703	26.2107	8.8237
688	473344	325660672	26.2298	8.8280
689	474721	327082769	26.2488	8.8323
690	476100	328509000	26.2679	8.8366
691	477481	329939371	26.2869	8.8408
692	478864	331373888	26.3059	8.8451
693	480249	332812557	26.3249	8.8493
694	481636	334255384	26.3439	8.8536
695	483025	335702375	26.3629	8.8578
696	484416	337153536	26.3818	8.8621
697	485809	338608873	26.4008	8.8663
698	487204	340068392	26.4197	8.8706
699	488601	341532099	26.4386	8.8748
700	490000	343000000	26.4575	8.8790
701	491401	344472101	26.4764	8.8833
702	492804	345948408	26.4953	8.8875
703	494209	347428927	26.5141	8.8917
704	495616	348913664	26.5330	8.8959
705	497025	350402625	26.5518	8.9001
706	498436	351895816	26.5707	8.9043
707	499849	353393243	26.5895	8.9085
708	501264	354894912	26.6083	8.9127
709	502681	356400829	26.6271	8.9169
710	504100	357911000	26.6458	8.9211

No.	Square	Cube	Square Root	Cube Root
711	505521	359425431	26.6646	8.9253
712	506944	360944128	26.6833	8.9295
713	508369	362467097	26.7021	8.9337
714	509796	363994344	26.7208	8.9378
715	511225	365525875	26.7395	8.9420
716	512656	367061696	26.7582	8.9462
717	514089	368601813	26.7769	8.9503
718	515524	370146232	26.7955	8.9545
719	516961	371694959	26.8142	8.9587
720	518400	373248000	26.8328	8.9628
721	519841	374805361	26.8514	8.9670
722	521284	376367048	26.8701	8.9711
723	522729	377933067	26.8887	8.9752
724	524176	379503424	26.9072	8.9794
725	525625	381078125	26.9258	8.9835
726	527076	382657176	26.9444	8.9876
727	528529	384240583	26.9629	8.9918
728	529984	385828352	26.9815	8.9959
729	531441	387420489	27.	9.
730	532900	389017000	27.0185	9.0041
731	534361	390617891	27.0370	9.0082
732	535824	392223168	27.0555	9.0123
733	537289	393832837	27.0740	9.0164
734	538756	395446904	27.0924	9.0205
735	540225	397065375	27.1109	9.0246
736	541696	398688256	27.1293	9.0287
737	543169	400315553	27.1477	9.0328
738	544644	401947272	27.1662	9.0369
739	546121	403583419	27.1846	9.0410
740	547600	405224000	27.2029	9.0450
741	549081	406869021	27.2213	9.0491
742	550564	408518488	27.2397	9.0532
743	552049	410172407	27.2580	9.0572
744	553536	411830784	27.2764	9.0613
745	555025	413493625	27.2947	9.0654
746	556516	415160936	27.3130	9.0694
747	558009	416832723	27.3313	9.0735
748	559504	418508992	27.3496	9.0775
749	561001	420189749	27.3679	9.0816
750	562500	421875000	27.3861	9.0856
751	564001	423564751	27.4044	9.0896
752	565504	425259008	27.4226	9.0937
753	567009	426957777	27.4408	9.0977
754	568516	428661064	27.4591	9.1017
755	570025	430368875	27.4773	9.1057

No.	Square	Cube	Square Root	Cube Root
756	571536	432081216	27.4955	9.1098
757	573049	433798093	27.5136	9.1138
758	574564	435519512	27.5318	9.1178
759	576081	437245479	27.5500	9.1218
760	577600	438976000	27.5681	9.1258
761	579121	440711081	27.5862	9.1298
762	580644	442450728	27.6043	9.1338
763	582169	444194947	27.6225	9.1378
764	583696	445943744	27.6405	9.1418
765	585225	447697125	27.6586	9.1458
766	586756	449455096	27.6767	9.1498
767	588289	451217663	27.6948	9.1537
768	589824	452984832	27.7128	9.1577
769	591361	454756609	27.7308	9.1617
770	592900	456533000	27.7489	9.1657
771	594441	458314011	27.7669	9.1696
772	595984	460099648	27.7849	9.1736
773	597529	461889917	27.8029	9.1775
774	599076	463684824	27.8209	9.1815
775	600625	465484375	27.8388	9.1855
776	602176	467288576	27.8568	9.1894
777	603729	469097433	27.8747	9.1933
778	605284	470910952	27.8927	9.1973
779	606841	472729139	27.9106	9.2012
780	608400	474552000	27.9285	9.2052
781	609961	476379541	27.9464	9.2091
782	611524	478211768	27.9643	9.2130
783	613089	480048687	27.9821	9.2170
784	614656	481890304	28.	9.2209
785	616225	483736625	28.0179	9.2248
786	617796	485587656	28.0357	9.2287
787	619369	487443403	28.0535	9.2326
788	620944	489303872	28.0713	9.2365
789	622521	491169069	28.0891	9.2404
790	624100	493039000	28.1069	9.2443
791	625681	494913671	28.1247	9.2482
792	627264	496793088	28.1425	9.2521
793	628849	498677257	28.1603	9.2560
794	630436	500566184	28.1780	9.2599
795	632025	502459875	28.1957	9.2638
796	633616	504358336	28.2135	9.2677
797	635209	506261573	28.2312	9.2716
798	636804	508169592	28.2489	9.2754
799	638401	510082399	28.2666	9.2793
800	640000	512000000	28.2843	9.2832

No.	Square	Cube	Square Root	Cube Root
801	641601	513922401	28.3019	9.2870
802	643204	515849608	28.3196	9.2909
803	644809	517781627	28.3373	9.2948
804	646416	519718464	28.3549	9.2986
805	648025	521660125	28.3725	9.3025
806	649636	523606616	28.3901	9.3063
807	651249	525557943	28.4077	9.3102
808	652864	527514112	28.4253	9.3140
809	654481	529475129	28.4429	9.3179
810	656100	531441000	28.4605	9.3217
811	657721	533411731	28.4781	9.3255
812	659344	535387328	28.4956	9.3294
813	660969	537367797	28.5132	9.3332
814	662596	539353144	28.5307	9.3370
815	664225	541343375	28.5482	9.3408
816	665856	543338496	28.5657	9.3447
817	667489	545338513	28.5832	9.3485
818	669124	547343432	28.6007	9.3523
819	670761	549353259	28.6182	9.3561
820	672400	551368000	28.6356	9.3599
821	674041	553387661	28.6531	9.3637
822	675684	555412248	28.6705	9.3675
823	677329	557441767	28.6880	9.3713
824	678976	559476224	28.7054	9.3751
825	680625	561515625	28.7228	9.3789
826	682276	563559976	28.7402	9.3827
827	683929	565609283	28.7576	9.3865
828	685584	567663552	28.7750	9.3902
829	687241	569722789	28.7924	9.3940
830	688900	571787000	28.8097	9.3978
831	690561	573856191	28.8271	9.4016
832	692224	575930368	28.8444	9.4053
833	693889	578009537	28.8617	9.4091
834	695556	580093704	28.8791	9.4129
835	697225	582182875	28.8964	9.4166
836	698896	584277056	28.9137	9.4204
837	700569	586376253	28.9310	9.4241
838	702244	588480472	28.9482	9.4279
839	703921	590589719	28.9655	9.4316
840	705600	592704000	28.9828	9.4354
841	707281	594823321	29.	9.4391
842	708964	596947688	29.0172	9.4429
843	710649	599077107	29.0345	9.4466
844	712336	601211584	29.0517	9.4503
845	714025	603351125	29.0689	9.4541

No.	Square	Cube	Square Root	Cube Root
846	715716	605495736	29.0861	9.4578
847	717409	607645423	29.1033	9.4615
848	719104	609800192	29.1204	9.4652
849	720801	611960049	29.1376	9.4690
850	722500	614125000	29.1548	9.4727
851	724201	616295051	29.1719	9.4764
852	725904	618470208	29.1980	9.4801
853	727609	620650477	29.2062	9.4838
854	729316	622835864	29.2233	9.4875
855	731025	625026375	29.2404	9.4912
856	732736	627222016	29.2575	9.4949
857	734339	629422793	29.2746	9.4986
858	736164	631628712	29.2916	9.5023
859	737881	633839779	29.3087	9.5060
860	739600	636056000	29.3258	9.5097
861	741321	638277381	29.3428	9.5134
862	743044	640503928	29.3598	9.5171
863	744769	642735647	29.3769	9.5207
864	746496	644972544	29.3939	9.5244
865	748225	647214625	29.4109	9.5281
866	749956	649461896	29.4279	9.5317
867	751689	651714363	29.4449	9.5354
868	753424	653972032	29.4618	9.5391
869	755161	656234909	29.4788	9.5427
870	756900	658503000	29.4958	9.5464
871	758641	660776311	29.5127	9.5501
872	760384	663054848	29.5296	9.5537
873	762129	665338617	29.5466	9.5574
874	763876	667627624	29.5635	9.5610
875	765625	669921875	29.5804	9.5647
876	767376	672221376	29.5973	9.5683
877	769129	674526133	29.6142	9.5719
878	770884	676836152	29.6311	9.5756
879	772641	679151439	29.6479	9.5792
880	774400	681472000	29.6648	9.5828
881	776161	683797841	29.6816	9.5865
882	777924	686128968	29.6985	9.5901
883	779689	688465387	29.7153	9.5937
884	781456	690807104	29.7321	9.5973
885	783225	693154125	29.7489	9.6010
886	784996	695506456	29.7658	9.6046
887	786769	697864103	29.7825	9.6082
888	788544	700227072	29.7993	9.6118
889	790321	702595369	29.8161	9.6154
890	792100	704969000	29.8329	9.6190

No.	Square	Cube	Square Root	Cube Root
891	793881	707347971	29.8496	9.6226
892	795664	709732288	29.8664	9.6262
893	797449	712121957	29.8831	9.6298
894	799236	714516984	29.8998	9.6334
895	801025	716917375	29.9166	9.6370
896	802816	719323136	29.9333	9.6406
897	804609	721734273	29.9500	9.6442
898	806404	724150792	29.9666	9.6477
899	808201	726572699	29.9833	9.6513
900	810000	729000000	30.	9.6549
901	811801	731432701	30.0167	9.6585
902	813604	733870808	30.0333	9.6620
903	815409	736314327	30.0500	9.6656
904	817216	738763264	30.0666	9.6692
905	819025	741217625	30.0832	9.6727
906	820836	743677416	30.0998	9.6763
907	822649	746142643	30.1164	9.6799
908	824464	748613312	30.1330	9.6834
909	826281	751089429	30.1496	9.6870
910	828100	753571000	30.1662	9.6905
911	829921	756058031	30.1828	9.6941
912	831744	758550528	30.1993	9.6976
913	833569	761048497	30.2159	9.7012
914	835396	763551944	30.2324	9.7047
915	837225	766060875	30.2490	9.7082
916	839056	768575296	30.2655	9.7118
917	840889	771095213	30.2820	9.7153
918	842724	773620632	30.2985	9.7188
919	844561	776151559	30.3150	9.7224
920	846400	778688000	30.3315	9.7259
921	848241	781229961	30.3480	9.7294
922	850084	783777448	30.3645	9.7329
923	851929	786330467	30.3809	9.7364
924	853776	788889024	30.3974	9.7400
925	855625	791453125	30.4138	9.7435
926	857476	794022776	30.4302	9.7470
927	859329	796597983	30.4467	9.7505
928	861184	799178752	30.4631	9.7540
929	863041	801765089	30.4795	9.7575
930	864900	804357000	30.4959	9.7610
931	866761	806954491	30.5123	9.7645
932	868624	809557568	30.5287	9.7680
933	870489	812166237	30.5450	9.7715
934	872356	814780504	30.5614	9.7750
935	874225	817400375	30.5778	9.7785

No.	Square	Cube	Square Root	Cube Root
936	876096	820025856	30.5941	9.7819
937	877969	822656953	30.6105	9.7854
938	879844	825293672	30.6268	9.7889
939	881721	827936019	30.6431	9.7924
940	883600	830584000	30.6594	9.7959
941	885481	833237621	30.6757	9.7993
942	887364	835896888	30.6920	9.8028
943	889249	838561807	30.7083	9.8063
944	891136	841232384	30.7246	9.8097
945	893025	843908625	30.7409	9.8132
946	894916	846590536	30.7571	9.8167
947	896809	849278123	30.7734	9.8201
948	898704	851971392	30.7896	9.8236
949	900601	854670349	30.8058	9.8270
950	902500	857375000	30.8221	9.8305
951	904401	860085351	30.8383	9.8339
952	906304	862801408	30.8545	9.8374
953	908209	865523177	30.8707	9.8408
954	910116	868250664	30.8869	9.8443
955	912025	870983875	30.9031	9.8477
956	913936	873722816	30.9192	9.8511
957	915849	876467493	30.9354	9.8546
958	917764	879217912	30.9516	9.8580
959	919681	881974079	30.9677	9.8614
960	921600	884736000	30.9839	9.8648
961	923521	887503681	31.	9.8683
962	925444	890277128	31.0161	9.8717
963	927369	893056347	31.0322	9.8751
964	929296	895841344	31.0483	9.8785
965	931225	898632125	31.0644	9.8819
966	933156	901428696	31.0805	9.8854
967	935089	904231063	31.0966	9.8888
968	937024	907039232	31.1127	9.8922
969	938961	909853209	31.1288	9.8956
970	940900	912673000	31.1448	9.8990
971	942841	915498611	31.1609	9.9024
972	944784	918330048	31.1769	9.9058
973	946729	921167317	31.1929	9.9092
974	948676	924010424	31.2090	9.9126
975	950625	926859375	31.2250	9.9160
976	952576	929714176	31.2410	9.9194
977	954529	932574833	31.2570	9.9227
978	956484	935441352	31.2730	9.9261
979	958441	938313739	31.2890	9.9295
980	960400	941192000	31.3050	9.9329

No.	Square	Cube	Square Root	Cube Root
981	962361	944076141	31.3209	9.9363
982	964324	946966168	31.3369	9.9396
983	966289	949862087	31.3528	9.9430
984	968256	952763904	31.3688	9.9464
985	970225	955671625	31.3847	9.9497
986	972196	958585256	31.4006	9.9531
987	974169	961504803	31.4166	9.9565
988	976144	964430272	31.4325	9.9598
989	978121	967361669	31.4484	9.9632
990	980100	970299000	31.4643	9.9666
991	982081	973242271	31.4802	9.9699
992	984064	976191488	31.4960	9.9733
993	986049	979146657	31.5119	9.9766
994	988036	982107784	31.5278	9.9800
995	990025	985074875	31.5436	9.9833
996	992016	988047936	31.5595	9.9866
997	994009	991026973	31.5753	9.9900
998	996004	994011992	31.5911	9.9933
999	998001	997002999	31.6070	9.9967
1000	1000000	1000000000	31.6228	10

Tables of squares, cubes, square root, and cube root will help many times in the solving of problems that occur in the plumbing trade. A great many pipe areas are determined by squaring the diameter, multiplied by .7854. Cubes are used often in figuring volumes. Reference to such tables will be found extremely helpful when decimals are used.

Fractions may be converted to decimals and as such the preceding tables may be used when figuring pipe capacities. When extreme accuracy is required, it will be found that ready reference to a square and cube root table will have a great many advantages.

WATER PRESSURE FOR VARYING HEADS

Head feet	Pressure lbs. per sq. inch	Head feet	Pressure lbs. per sq. inch	Head feet	Pressure lbs. per sq. inch
1	0.43	35	15.16	69	29.88
2	0.86	36	15.69	70	30.32
3	1.30	37	16.02	71	30.75
4	1.73	38	16.45	72	31.18
5	2.16	39	16.89	73	31.62
6	2.59	40	17.32	74	32.05
7	3.03	41	17.75	75	32.48
8	3.46	42	18.19	76	32.92
9	3.89	43	18.62	77	33.35
10	4.33	44	19.05	78	33.78
11	4.76	45	19.49	79	34.21
12	5.20	46	19.92	80	34.65
13	5.63	45	20.35	81	35.08
14	6.06	48	20.79	82	35.42
15	6.49	49	21.22	83	35.95
16	6.93	50	21.65	84	36.39
17	7.36	51	22.09	85	36.82
18	7.79	52	22.52	86	37.25
19	8.22	53	22.95	87	37.68
20	8.66	54	23.39	88	38.12
21	9.09	55	23.82	89	38.55
22	9.53	56	24.26	90	38.98
23	9.96	57	24.69	91	39.42
24	10.39	58	25.12	92	39.95
25	10.82	59	25.55	93	40.28
26	11.26	60	25.99	94	40.72
27	11.89	61	26.42	95	41.15
28	12.12	62	26.89	96	41.58
29	12.55	63	27.29	97	42.01
30	12.99	64	27.72	98	42.45
31	13.42	65	28.15	99	42.88
32	13.86	66	28.58	100	43.31
33	14.29	67	29.02	200	86.62
34	14.72	68	29.45	300	129.93

If pressure is known; to find head feet multiply by 2.309 or 2.31

MEASURES USED IN PIPING PROBLEMS

1 U.S. gallon weighs = 8.33 pounds.

1 cubic foot water (fresh) = 62.4 pounds.

1 cubic foot sea water = 64.1 pounds.

1 cubic foot = 7.48 U.S. gallons.

1 U.S. gallon = 0.1337 cubic foot.

1 U.S. gallon = 231 cubic inches.

1 British Imperial gallon of water = 10 pounds.

1 pound per square inch pressure depth of water = 2.309 or 2.31.

1 foot depth of water = 0.433 pounds per square inch.

1 atmosphere, atmospheric pressure = 14.7 pounds per square inch.

1 atmosphere = 33.9 feet of water.

1 atmosphere = 29.9 inches of mercury.

Specific gravity of Mercury = 13.6.

Acceleration of gravity (g) = 32.2 feet per second.

Square root of acceleration symbol $\sqrt{29}$ = 8.02.

1 cubic foot per second = 449 gallons per minute.

1 acre foot = 43,560 cubic feet.

1 horse power = 1 cubic foot of water per second, × 8.81 feet of head.

If pressure is known and it is desired to find head, multiply the pressure by 2.309 or 2.31.

If head is known and it is desired to find pressure, multiply head by .433.

psi means pounds per square inch.

gpm means gallons per minute.

TABLE OF MEASURES

```
Measures of time

60 seconds ---------------------------------------------------1 minute
60 minutes ---------------------------------------------------1 hour
24 hours ------------------------------------------------------1 day
7 days --------------------------------------------------------1 week
365 days or 52 weeks -------------------------------------- 1 year
 12 months --------------------------------------------------1 year
 366 days ----------------------------------------------------- 1 leap year
 100 years ----------------------------------------------------- 1 century
```

```
Metric table of capacity

10 milliliters -----------------------------------------------1 centiliter
10 centiliters -----------------------------------------------1 deciliter
10 deciliters ------------------------------------------------1 liter----or
                                                            (1.057 quarts)
10 liters -----------------------------------------------------1 dekaliter
10 dekaliters -----------------------------------------------1 hectoliter
10 hectoliters ----------------------------------------------1 kiloliter--or
                                                            (1.308 cu. yds.)
```

```
Metric table of weight

10 milligrams ----------------------------------------------1 centigram
10 centigrams ----------------------------------------------1 decigram
10 decigrams -----------------------------------------------1 gram
10 grams ----------------------------------------------------1 dekagram
10 dekagrams ----------------------------------------------1 hectogram
10 hectograms ---------------------------------------------1 kilogram---or
                                                            (2.205 pounds)
```

```
Circular measure

60 seconds      (")-----------------------------------1 minute (')
60 minutes      (')-----------------------------------1 degree (°)
90° degrees     (°)-----------------------------------1 quadrant
360° degrees    (°)-----------------------------------1 circumference
```

Surveyors linear measure

7.92 inches ---1 link
25 links ---1 rod
4 rods ---1 chain
80 chains ---1 mile

Dry measure

2 pints ---1 quart
8 quarts ---1 peck
4 pecks ---1 bushel

Liquid measure

4 gills ---1 pint
2 pints ---1 quart
4 quarts ---1 gallon
31½ gallons ---1 barrel
2 barrels or 63 gallons ---1 hogshead

Avoirdupois weight

16 ounces ---1 pound
100 pounds ---1 hundred wt.
20 hundredweights ---1 ton (short)
2240 pounds ---1 long ton

Measures of money

10 mills ---1 cent
10 cents ---1 dime
10 dimes ---1 dollar
10 dollars ---1 eagle

Stationer's measure
24 sheets ---1 quire
20 quires ---1 ream
500 sheets ---1 ream

Linear measure

12 inches	1 foot
3 feet	1 yard
5½ yards	1 rod
40 rods	1 furlong
8 furlongs	1 mile
1 mile	5280 feet
3 miles	1 league

Square measure

144 square inches	1 square foot
9 square feet	1 square yard
30¼ square yards	1 square rod
160 square rods	1 acre
1 acre	43,560 sq. feet
640 acres	1 square mile

Cubic measure

1728 cubic inches	1 cubic foot
27 cubic feet	1 cubic yard
128 cubic feet	1 cord (wood)
2150.42 cubic inches	1 standard bushel
2688 cubic inches	1 heaped bushel
231 cubic inches	1 U.S. standard gal.

Metric table of length

10 millimeters	1 centimeter
10 centimeters	1 decimeter
10 decimeters	1 meter (39.37 in.)
10 meters	1 dekameter
10 dekameters	1 hectometer
10 hectometers	1 kilometer (0.62 mi)

VELOCITIES THEORETICALLY GIVEN TO FLOW DUE TO HEAD FEET

Head Ft.	Vel. Ft. per sec.	Head Ft.	Vel. Ft. per sec.	Head Ft.	Vel. Ft. per sec.	Head Ft.	Vel. Ft. per sec.
1 Ft.	8.03	3.5	15.	19.	35.	66.	65.2
		3.6	15.2	.5	35.4	67.	65.7
		3.7	15.4	20.	35.9	68.	66.2
		3.8	15.6	.5	36.3	69.	66.7
		3.9	15.8	21.	36.8	70.	67.1
1.10	8.41	4.	16.	.5	37.2	71.	67.6
		.2	16.4	22.	37.6	72.	68.1
		.4	16.8	.5	38.1	73.	68.5
		.6	17.2	23.	38.5	74.	69.
		.8	17.6	.5	38.9	75.	69.5
1.20	8.79	5.	17.9	24.	39.3	76.	69.9
		.2	18.3	.5	39.7	77.	70.4
		.4	18.7	25.	40.1	78.	70.9
		.6	19.	26.	40.9	79.	71.3
		.8	19.3	27.	41.7	80.	71.8
1.30	9.15	6.	19.7	28.	42.5	81.	72.2
		.2	20.	29.	43.2	82.	72.6
		.4	20.3	30.	43.9	83.	73.1
		.6	20.6	31.	44.7	84.	73.5
		.8	20.9	32.	45.4	85.	74.
1.40	9.49	7.	21.2	33.	46.1	86.	74.4
		.2	21.5	34.	46.7	87.	74.8
		.4	21.8	35.	47.4	88.	75.3
		.6	22.1	36.	48.1	89.	75.7
		.8	22.4	37.	48.8	90.	76.1
1.50	9.83	8.	22.7	38.	49.5	91.	76.5
		.2	23.	39.	50.1	92.	76.9
		.4	23.3	40.	50.7	93.	77.4
		.6	23.5	41.	51.3	94.	77.8
		.8	23.8	42.	52.	95.	78.2
1.60	10.2	9.	24.1	43.	52.6	96.	78.6
		.2	24.3	44.	53.2	97.	79.
		.4	24.6	45.	53.8	98.	79.4
		.6	24.8	46.	54.4	99.	79.8
1.80	10.8	.8	25.1	47.	55.	100.	80.3
		10.	25.4	48.	55.6	125.	89.7
1.90	11.1	.5	26.	49.	56.2	150.	98.3
		11.	26.6	50.	56.7	175.	106.
2.	11.4	.5	27.2	51.	57.3	200.	114.
2.1	11.7	12.	27.8	52.	57.8	225.	120.
2.2	11.9	.5	28.4	53.	58.4	250.	126.
2.3	12.2	13.	28.9	54.	59.	275.	133.
2.4	12.4	.5	29.5	55.	59.5	300.	139.
2.5	12.6	14.	30.	56.	60.	350.	150.
2.6	12.9	.5	30.5	57.	60.6	400.	160.
2.7	13.2	15.	31.1	58.	61.1	450.	170.
2.8	13.4	.5	31.6	59.	61.6	500.	179.
2.9	13.7	16.	32.1	60.	62.1	550.	188.
3.	13.9	.5	32.6	61.	62.7	600.	197.
3.1	14.1	17.	33.1	62.	63.2	700	212.
3.2	14.3	.5	33.6	63.	63.7	800	227.
3.3	14.5	18.	34.	64.	64.2	900.	241.
3.4	14.8	.5	34.5	65.	64.7	1000.	254.

VELOCITIES FOR GIVEN SLOPES

Approximate velocities for given slopes and diameters for rough pipe. These tables may be used to determine a building drain or sewer that will ensure a velocity above some selected minimum. Check before installing.

Diameter of pipe	Velocities in feet per second according to fall.				
	1/32"	1/16"	1/8"	1/4"	1/2"
1 "	0.57	0.80	1.14	1.61	-2.28
1½"	.62	.88	1.24	1.76	-2.45
2 "	.72	1.02	1.44	2.03	-2.88
2½"	.81	1.14	1.61	2.28	-3.23
3 "	.88	1.24	1.76	2.49	3.53
4 "	1.02	1.44	2.03	2.88	-4.07
5 "	1.14	1.61	2.28	3.23	-4.56
6 "	1.24	1.76	2.49	3.53	-5.00
8 "	1.44	2.03	2.88	4.07	-5.75
10 "	1.61	2.28	3.23	4.56	-6.44
12 "	1.76	2.49	3.53	5.00	-7.06

Fractions denote the pitch in inches per foot.
Velocity in feet per second below pitch shown.

For long runs of pipe where grade is a factor, it may be necessary to increase the pipe size to compensate for the lack of grade. In order to maintain a velocity that will ensure a proper flow, slope and diameter must be considered. The table above may be used in considering long building drains or sewers. Ordinary fall to secure at least 2 feet per second is shown in the table above.

Check with the local inspection department on sewers or drains when minimum allowed grade cannot be obtained.

VENTING CAPACITIES OF FLUES

Natural Draft Venting Conditions. Forced draft of exhaust systems require special engineering. Check for approval requirements. In sizing flues or vents, make sure that the vent is adequate. When more than one appliance vents into a flue or vent, the flue or vent area must not be less than the area of the largest flue or vent connector plus 50 percent of the areas of the additional flue or vent connectors.

Ratings and diameters			
Maximum Input B.T.U. rating (B.T.U.)	Diameter of pipe (inches)	Area of pipe (sq. inches)	½ Area of pipe (sq. in)
65,000	4	12.56	6.28
100,000	5	19.63	9.88
150,000	6	28.27	14.14
200,000	7	38.48	19.24
300,000	8	50.26	25.13
380,000	9	63.60	31.80
480,000	10	78.54	39.27
580,000	11	95.03	47.51
680,000	12	113.09	56.54
930,000	14	153.93	76.96
1,200,000	16	201.06	100.53

Always check for the approval and stamp of the recognized testing agency for Btu rating and vent clearances. Stack temperatures on large size installations may require special fireproofing. Check for special conditions on water heaters and manifold requirements.

Never combine flues or vents until all special requirements are known. Where excessive lengths are encountered, check with the local inspection department. (See water heater manifold page 237.) Check local ordinance before combining as shown.

PROFESSIONAL PLUMBING ILLUSTRATED EXAMINATIONS

A modern approach to an old problem of presenting questions and answers that are fully understood by the applicant.

Code Examinations. 100 illustrated examples with multiple choice answers. Problems deal mainly with "on the job" plumbing design and installation knowledge required by code.

Trade Practice. 100 illustrated examples with multiple choice answers. Problems deal with math, tools, installation and general "trade" knowledge. All material is strictly "up to date" and is based on uniform practice and national plumbing procedure. Adapts to all sections.

1. Presented in a professional-type 8-by-10 booklet of 10 pages with 10 illustrations on each page. Front page for applicant registrations and examination explanation. Rear page for examiners use and records.

2. Key Corrected. Key furnished with each examination, easy to use and completely accurate.

3. Examination Time. Two hours for 100 questions or approximately 1¼ minutes to a question.

4. Correction Time. Two to three minutes.

Next to the actual installation, a picture is the best method of communication. This is the principle used in these exams. "See the picture" and select the answer. (Wrong answers may be checked and associated with the illustrations, once understood, never forgotten.)

Used in conjunction they are an unbeatable combination. Applicants scoring a passing grade would be considered a competent, qualified, first-class journeyman in any section of the nation.

....

PROFESSIONAL PLUMBING ILLUSTRATED EXAMINATIONS

Please send _____ "CODE EXAMINATIONS" to:
Please send _____ "TRADE PRACTICE EXAMINATIONS" to:
Name _____
Address _____
City _____
State _____ Zip code _____

Examination prices ... $3.00 each
(Add 6% sales tax in California)

PUBLISHER ....DISTRIBUTOR

A. J. Smith Prof. Plumb.
GORDON LOYD ENTERPRISES
P.O. Box 3428 (209) 733-5706
Visalia, CA. 93278

....

PROFESSIONAL PLUMBING ILLUSTRATED EXAMINATIONS

Please send _____ "CODE EXAMINATIONS" to:
Please send _____ "TRADE PRACTICE EXAMINATIONS" to:
Name _____
Address _____
City _____
State _____ Zip code _____

Examination prices $3.00 each
(Add 6% sales tax in California)

PUBLISHER   DISTRIBUTOR

A. J. Smith Prof. Plumb.
GORDON LOYD ENTERPRISES
P.O. Box 3428 (209) 733-5706
Visalia, CA. 93278

OTHER POPULAR TAB BOOKS OF INTEREST

TAB   TAB BOOKS Inc.

Blue Ridge Summit, Pa. 17214

Send for FREE TAB Catalog describing over 750 current titles in print.